Integrated Data Warehouse for Indian Agriculture

INTEGRATED DATA WAREHOUSE FOR INDIAN AGRICULTURE

Dr. Aditya Kumar Gupta
Associate Professor, School of Management Sciences, Varanasi

NAMYA PRESS.com

Published by

Namya Press

213, Vardhan house, 7/28, Ansari Road, Darya Ganj, Delhi - 110002.

Email: namyapress@gmail.com

Edition : First Published in 2021 (Namya Press)

Title : Integrated Data Warehouse for Indian Agriculture

Author : Dr Aditya K Gupta

ISBN : 978-93-90445-12-7

Prof. P. N. Jha
D. Litt. (Management)
(Former Director, JDIMT, Punjab)
DIRECTOR

Accredited 'A' Grade by NAAC
SMS VARANASI SCHOOL OF MANAGEMENT SCIENCES
Approved by : AICTE Ministry of HRD, Govt. of INDIA

FOREWORD

It gave me immense pleasure and utmost satisfaction to see a manuscript entitled, 'Integrated Data Warehouse for Indian Agriculture', authored by my colleague, Dr. Aditya Kumar Gupta, Associate Professor. Following a sincere perusal of his present erudition, I feel the creative work represents a rich intellectual treasure in itself that has been elegantly written and carries a deft discussion in a lucid manner. Meticulously researched, the tangible creation attempts to meaningfully preserve and enrich the academic treasure pertaining to relevant data and information towards the currently turbulent sector of Indian Agriculture.

It has been a well-known proposition that agriculture-sector is the mainstay of the Indian economy, as it contributes around 18% of total GDP of the country, and about half of India's population is wholly or significantly dependent on agriculture & allied activities for their livelihood. It is equally an open truth that India is among the 15 leading exporters of agricultural products in the world, and has further the 10th largest arable land resources globally. In an accelerated pace of digital era, application of ICT, also in agriculture, is increasingly important. The Electronic National Agriculture Market (eNAM) has already been launched. E-Agriculture is an emerging field focusing on the enhancement of agricultural and rural development through improved information and communication processes. More and more farmers are coming forward to understand how leveraging solutions that use the latest technologies such as Artificial Intelligence, Machine Learning, and Cloud offer them greater climate resilience, higher crop yield, and better price control. In view of enabling the agriculture-sector to effectively contribute towards the economic reforms in the country, and also to actualize the ambitious goal of the Government of India of doubling farm income by 2022, it looks imperative that all relevant stakeholders should have easier and better access to the information & data on latest varieties, changing weather patterns, crop production techniques and improved agronomic practices for managing various related activities more efficiently. Data warehousing is of course one of the important concepts that is capable enough to provide the needful solutions.

I feel inclined to hereby extend my heartfelt greetings to Dr. Gupta for undertaking a trailblazing and inter-disciplinary academic exercise. I am sure the book would be widely consulted.

9/2/2021

(P. N. Jha)

CAMPUS : KHUSHIPUR (MOHANSARAI-MUGHALSARAI BYPASS), P.O. - BACHHAON, VARANASI - 221011 • **Phone**: 7052055555, 8953761666, 8953940777
CITY OFFICE : 6th FLOOR, VINAYAK PLAZA, MALDAHIYA CROSSING, VARANASI - 221002 • **Phone** : 8400277770
E-mail : info@smsvaranasi.com • **Visit us** : www.smsvaranasi.com

भारतीय प्रबन्ध संस्थान
प्रबन्ध नगर, ऑफ सीतापुर रोड,
लखनऊ-226 013 (उ.प्र.), भारत

Indian Institute of Management
Prabandh Nagar, Off Sitapur Road,
Lucknow-226 013 (U.P.) India

Prof. K. B. Gupta

Centre for Food & Agribusiness Management

Agriculture is one of the most important sectors of the Indian economy and requires technical breakthrough to achieve its optimum potential. In today's competitive ecommerce world, agricultural data processing is an increasing need of Indian farmers. The application of information technology in the field of agriculture is instrumental in making appropriate decisions and responding to queries of the farmers in an appropriate manner.

I congratulate the author of the book as his effort would really be very useful and help in leveraging the data warehouse technology for providing real life solutions to frame the short-term as well as the long-term strategies for agriculture and its allied sectors.

Best Wishes,

(Prof. K.B. Gupta)

Institute of Engineering and Rural Technology, Allahabad

Department of Computer Science and Engineering

Engineering Degree Division

26, Chatham line, Allahabad - 211002, UP, INDIA

Dr. Bireshwar Dass Mazumdar

Associate Professor & Head

Contact: (+91)-532-2544810

(+91)-9307977683

bdmazumdar@iert.ac.in

February 10, 2021

The Radical objective of this work is to create database about Indian crops and their requirements. This work provides a detail framework to find solution of problems for Indian farmers and gives response of their Ad-hoc quires. The key objective of proposed Data warehouse is to helps farmers for planning and forecasting about agricultural & environmental issues. Although it seems a broad scope, which cannot be covered in a single research work area, but I hope this is a good deal of Information Technology in area of agriculture. I am sure that the work done by the author will promote other researchers to add their contribution in the field of agriculture.

I am extremely delighted to write the forewords of this research work, which is carried out under my supervision. I truly appreciate the enthusiasm, passion and hard work of the author and I enjoyed to work with him.

Best Wishes and Good Luck!

(Dr. B.D. Mazumdar)

Preface

Agriculture is one of the most important industry of India that contributes around 18 percent of total GDP of the nation. Approximately 60 percent of Indian population is depending on the agricultural industry for their livelihood. The Indian National Anthem also articulates the names of all agricultural dominated regions of India. The very first stanza of Indian National Anthem also includes the name of Indian states "Punjab, Sindh, Gujarat, Maharashtra, Tamil-Nadu, Odisha, Bengal, Vindhayas, The Himalayas and regions blessed by river "Ganges and Yamuna" these all regions plays a significant role in the Indian Agriculture industry.

Human civilization has been growing crops and raising livestock for approximately 11,000 years. Throughout this period, farmers have continually adapted their technologies, assessed the results, and shared what they have learned with other members of the community. Most of this communication has taken the form of verbal explanations and practical demonstrations, but some information took a more durable form as soon as systems of writing were developed. Details of agricultural practices have been found in records from ancient Egypt, Mesopotamia and China going back more than 3,000 years. It was not until the beginning of the 20th century, when colleges in the United States started conducting demonstrations at agricultural shows and giving lectures to farmer's clubs. Since then, many research initiatives have been taken in the field of agricultural technology. However, applications of Information Technology and Computer Sciences in agriculture sector are still in infancy.

The main objective of this research work is to empower the agriculture sector with the new advancement of Information Technology and related areas. Data warehousing is one of the important concepts that is capable enough to provide solutions of Ad-hoc queries of farmers for efficient planning of cropping system. Since last one decade the data warehouses have been implemented in the field of customer relationship, marketing and finance. In response to advent of new technologies researchers also start to apply knowledge of information technology in field of agriculture. This research work led to design an "Integrated Data warehouse for Indian Agriculture".

Acknowledgements

After completing this work, it gives me great pleasure to express my gratitude to Omnipresent God "Shri Raj ji" and Goddess "Saraswati ji ". It gives me pleasure in placing one record to my greatest sense of gratitude to my first teacher who made me able to understand the set of alphabets and gave me intelligence to convert them into meaningful sentences. I do not have words to express my thanks to my parents and my brothers. They motivated me and provide moral support and enthusiasm throughout my career. I feel myself obelised towards my parents and both of my beloved brothers.

I am sincerely thankful, to my supervisor, and truly a guide, Dr. Bireshwar Dass Mazumdar, Associate Professor, IERT Prayagraj for his, expertise knowledge and support. I pay many thanks to him, and I really mean it.

I would like offer my millions thanks to Professor P.N. Jha, Director, School of Management Sciences-Varanasi, for his blessings and enthusiastic support. I also offer my gratitude to my co-workers and friends for their time-to-time advice and help. A special thanks to Professor K. S. Mishra and Dr. Ravi Shekhar Vishal for their encouragement and vigilance reviews.

During the long journey of my research, I have received lot of ideas, thoughts and suggestions from many researchers and academicians. I wish to pay sincere thanks to Professor Kriti Bardhan Gupta- IIM Lucknow, Professor Jabir Ali- IIM

Lucknow, Dr. D.S. Kushwaha-I.E.T. Lucknow, Dr. M.H. Khan-I.E.T. Lucknow. The names are un-numeral.

This acknowledgment will be incomplete if I fail to express a sense of gratitude to my wife 'Chakit' for her technical support while editing this book. During preparation of this book, I also acknowledge a sense of care towards my lovely children Bhvya and Dakshita, who has many times waited for me to enjoy family time.

Although I have tried to express my gratitude towards every person who contributed to my work, there may someone still hiding veils of unknown, whom I may not have been able to recognize. I would like to thank all of them who have worked for me without my knowledge.

Aditya Kumar Gupta

List of Tables

List of Figures

Contents

CHAPTER 1

Introduction

Agriculture is one of the most important inventions of human civilization. The development of human civilization and development of agriculture technology were the two wheels of the cart. Unfortunately, it has been witnessed that the development of agriculture technology is not in the same ratio as human civilization is developed.

The evidence of proper of agriculture practices in Indian sub-continent is reported since Indus Valley civilization era and even before that in some parts of south regions of present India. Around 9000 BCE some north-west regions of present India started the early cultivation of plants, and domestication of useful animals. In the result, the life of early human in India is started to get settled. Around 8000-4000 BCE during Neolithic revolution the practices of Agro pastoralism is stated and the primitive techniques being developed for agriculture. Around 4500 BCE, the human of Indus Valley Civilisation stated proper Irrigation and they have required knowledge about session and importance of rivers for irrigation. They have developed a calendar for cropping systems and was aware of rain-cycle and double-rainfall in specific regions of prehistoric India. Archaeological evidences are also available to articulate presence of sophisticated irrigation system and presence of artificial reservoirs around 3000 BCE. The innovative use of irrigation is also reported by Indus civilisation in form of well-planned drainage and sewage system that logically proves that they use to use artificial canals as primitive irrigation solutions.

The Vedic literature provides repeated references of cultivation of barley, wheat, vegetables and fruits. The agricultural technology between 1500 BCE and 200 CE includes animal husbandry, ploughing the soil, broadcast seeds, and used a certain sequence of cropping for better production.

Around 322–185 BCE, During the Mauryan Empire the construction and maintenance of dams are reported. The meteorological observations of soils are in practice and soil category is checked for particular crops. The Greek visitor Megasthenes (300 BC) has reported the same in his book *Indika* that provides an interesting eyewitness account of Indian agriculture.

During early Common Era to high middle ages around 200–1200 CE, the agriculture advancement is also started in southern and eastern regions of India. The Dravidian people has stated cultivation of grain like rice, beans, coconut, sugarcane, millets, black pepper, tamarind and sandalwood. They were practising of systematic ploughing, manuring, weeding, irrigation and crop protection for sustained agriculture. The Water storage systems were designed during this period, and it is interesting to know that Kallanai dam is an ancient dam built around 1st-2nd century CE across the Kaveri river between Tiruchirapalli District and Thanjavur district, is considered to be one of the oldest water-regulation structures in the world that is still in use.

Between 875-1279 during the Chola Empire in South India the land was transferred and hold in possession that indicates ownership of land plots to the individual civilian, that may be consider the early evidence of individual disposition of farming property. Another noticeable advancement of during this period was having dedicated bureaucrats for proper monitoring and implementation of distribution of water particularly to the drier areas using water tanks and channel networks.

In the Late middle age (around 1200-1526) the impact of Persian irrigation technologies becomes reason of advancement of Indian irrigation practices. During this period many water reservoirs, water wells and artificial canals are constructed by local ruler, some of them are still available. Further, it is also reported that two agricultural zones were created during this period one indicating production of rice and another indication to production wheat or millets. Sugar mills was also appeared in India during this era.

It is considered that around 1540 at Delhi and nearby place there are few primitive sugar mills however the non-commercial practices for making sugar may date back too earlier and common in the northern Indian sub-continent. Geared sugar rolling mills later appeared in Mughal period. It was based on the principle of rollers as well as worm gearing. It was continued till the 17th century.

During Mughal period the agricultural production increased and due to accelerated population growth variety of crops were stated to cultivate, including grain crops such as wheat, rice, and barley, and cash crops such as cotton, opium and indigo. Later Indian farmers also started cultivation of new crops like maize and tobacco originated from America. However, the Indian crops; such as cotton, sugar, and fruits are also spread world-wide visibly to North Africa and Middle East. "*The History of Agriculture*" by Britannica Educational Publishing" provide the details about migration of many crops in India during this period.

> *Cultivation of tobacco, introduced by the Portuguese spread rapidly. The Malabār Coast was the home of spices, especially black pepper, that had stimulated the first European adventures in the East. Coffee had been imported from Abyssinia and became a popular beverage in aristocratic circles by the end of the century. Tea, which was to become the commoner's drink and a major export, was yet undiscovered, though it was growing wild in the hills of Assam. Vegetables were cultivated mainly in the vicinity of towns. New species of fruit, such as the pineapple, papaya, and cashew nut, also were introduced by the Portuguese. The quality of mango and citrus fruits was greatly improved.*
>
> *-The History of Agriculture*

During Akbar's empire between 1556 and 1605, Todarmal, the scholar-bureaucrat of Akbar has formulated elaborated methods for agricultural management on a rational basis. The Land management implemented by Todermal was significantly strong and presently in practice in many regions of India. The Akbar's bureaucrat had also introduced a new land revenue system called *zabt*. He had replaced the previous tribute system, with a new monetary tax system based on a uniform currency.

In the British regime the irrigation works was promoted but not on the scale required at that time. The British government paid only required attention on promotion of agricultural technology in India. It is interesting to know at the same time formers of European countries were using relatively advance agriculture technology. The British Government majorly concentrated on commercial crops such as Indigo, Opium, Cotton, Wheat and Rice in order to made profit from global market. During this period Community effort based on the private investments were made for improvement of irrigation system.

Besides the Indus Valley civilization, there are two regions with distinct agriculture practices are also reporting in around 2800-1500 BCE. These are the Deccan Plateau and the area within the modern states of Orissa and Bihar. The ashmound tradition was developed around 2800 BCE in these regions. The people of the ashmound tradition cultivate millets, pulses and other crops. Study reports many sites in the southern Indian states of Tamil Nadu, Andhra Pradesh and Karnataka providing clear evidence of agriculture of pulses and, millet-grasses, wheats, barley pearl millet, finger millet, cotton as well as gathered fruits. They were also involved in herding cattle like cow, sheep and goat and were largely engaged in pastoralism (reported by Fuller 2006, 'Dung mounds and Domesticators'). In the eastern part of India people also use to cultivate rice and pulses, as well as keeping cattle, sheep and goat. This was the time when plants and animals ware considered as essential to survival and also are supposed to be worshiped and cherish.

1.1 Initiative taken by the Republic of India

After 1947 the Republic of India stated many programmes to improve production of grain and commercial crops. The Grow More Food Campaign (अधिक अन्न उपजाओ अभियान), was launched in 1942 to promote crop production by demonstration and contact with the farmers. But soon the campaign failed to achieve its targets. And after independence in 1948, the Central Government re-defined the objectives of the Campaign and add the attainment of self-sufficiency in food grains by 1952. At the same time, arrangements were made for integration and co-ordination of the entire campaign for increasing agricultural production. Some state governments associated the public with working of the campaign by setting up non-official committees at the village, taluka, district and state levels.

In the year 1950 The Integrated Production Programme was launched aiming technical change in agriculture but it was majorly focused on supply of food and cash crops. However, the very first Five-year plan of India was introduced for agricultural development including: Land reclamation, land development, electrification of village, use of pesticides, chemicals and fertilisers.

Few 'production revolutions' were also initiated after 1960 included "Green Revolution" (197-1977)", "Yellow Revolution" (1986-1990), "Operation Flood" (1970-1996), and "Blue Revolution" (1973-2002). The New economic reforms were stated in year 1991, that resultant the significant growth in the agricultural sector, which was the first step for India to think newer innovations and involvement of Agro-processing and Biotechnology in Agricultural sector.

1.1.1 Agriculture Research and Engineering

In British India the Royal Commission on Agriculture presented a report in 1928 that laid stress on using science to develop and spread new agricultural technologies for the irrigation of, dry and semi-dry areas. However, the magnitude of efforts made in agricultural research and engineering is limited until independence. Research in agricultural engineering is consider to begin at the Allahabad Agricultural Institute, Naini in 1921. During 1930 agricultural engineering research was started at the Agricultural College and Research Institute, Coimbatore.

Later, Allahabad Agricultural Institute stated B.Sc. Agricultural Engineering Programme in 1942, following, establishment of Agricultural Engineering Department at IIT, Kharagpur in 1954, afterword many others initiatives have taken place and colleges of Agricultural Engineering and Technology at Ludhiana, Pantnagar, Udaipur, Jabalpur and Coimbatore were established. After that many research opportunities were opened up in the water and soil engineering with the establishment of the Damodar Valley Corporation in 1949, aiming to solve the problems of soil and water conservation in West Bengal and Bihar.

The Government of India stated other initiative for establishing soil conservation centres at different regions of the country provisioned in the First Five-Year plan. Subsequently, in 1975 all these centres having same objective are combined together as a Central Soil and Water Conservation Research Institute at Dehra Dun under the Indian Council of Agricultural Research (ICAR). Good to sound that organizations other than the ICAR have shown interest in sponsoring research in different areas of agricultural engineering by giving financial support or as integral part of the activity of these organizations. Ministry of Non-Conventional Energy Sources, Department of Electronics, Department of Science and Technology, Department of Agriculture and Co-operation, Tata Energy Research Institute and the Indian National Committee of Irrigation and Drainage are names of such organizations. Few more organizations such as the National Dairy Development Board was established in 1965, National Bank for Agriculture and Rural Development was established in 1982 that aided the formation of cooperatives and improved financing. The ICAR has however remained the most important organization in the country supporting the research in areas of agricultural engineering and technologies. In 1954 the ICAR for the first time sponsored a scheme to conduct state-wise survey of the existing tools and implements used by farmers. ICAR gives the following recommendations after the said survey.

Recommendations

1. The work of improvement, development and standardization of indigenous implements should be carried out in co-ordinated manner with due regard to different soils, climatic conditions and cultural practices prevailing in various regions of the country.
2. Standardization of components of approved agricultural implements should be done to facilitate their manufacture on a mass scale, to be made available to interested farmers readily and cheaply.
3. Multi-purpose implements or tool bars with suitable attachments covering a variety of operations, similar to the bullock drawn tool carrier-cum-farm cart should be developed.
4. Research should be undertaken on animal yokes and methods of hitching with a view to improve their working efficiency.
5. Field tests of the implements developed should be conducted systematically and reported on a standard proforma.
6. Promising types of foreign implements should also be tried out under various soil and climatic conditions in different regions with a view to evolve successful designs.

This research work is aiming to fulfilling few recommendations of ICAR that are remain un-answered till the writing of this manuscript. This work focuses to the point 2 & 5 of the recommendation.

1.2 Background and Motivation

The Government of India has been earnestly trying to put into operation different plans to increase investment or outlay in merchandizing and commercializing. Some of the known plans and strategies of the Indian Government include the following:

1. Market Research and Information Network
2. Construction of Rural Go-downs
3. Grading and Standardization of Agriculture product
4. Development/Strengthening of Agricultural Marketing Infrastructure

Building a Data Warehouse with decision making capabilities is the core idea for pursuing this work that leads to fulfil first point i.e., "Market Research and Information Network", additionally, the proposed data warehouse is also useful for standardization of components and systematically reporting with rapid and fast decision making. Keeping above in mind this research also focuses in construction of "Integrated Data Warehouse for Indian Agriculture".

1.3 Objective of the Research Work

The basic objective of this research work is to provide more efficient approaches for farmers and decision makers. The ultimate goal is not only constructing a data warehouse for agriculture but to develop a decision support system to response ad-hoc queries of the farmers. The proposed decision support system provides strategies and remedies to maximize the production of a crop. It also gives holistic view of agricultural attributes and help to understand the impact of these attributes on production of a particular crop. Another objective of the thesis is to incorporate agent technology and data mining techniques to retrieval of information and to store of the information in warehouse. The multi-agent-based data mining leads to self-learning approach of the data warehouse.

Categorically the following tasks have been systematically and in order taken to meet the objectives:

1. To perform an appropriate and exhaustive literature survey pertaining to the farming in India. The understandings of parameters that majorly affect the crop production are one of core objective of this research. The researchers and scientist regroup, would unable to identify the untouched problem of agriculture sector. In order to achieve high performance, the application of data warehouse technology in the area of agriculture may create interest for research prospective.

2. To provide a design for data warehouse for agriculture sector that is useful for farmers and decision makers in decision making. The aspects understand weather report, examination of soil type, estimation of seed quality etc. The major objective is to understand the "need and lack" of a particular attribute that are important for cropping system.

3. To develop a mathematical model for agricultural decision support system. The agricultural parameters soil type, fertilizers, seeding-month etc. The proposed mathematical model first computes the values and then stored it into the multidimensional schema of data warehouse. This approach reduces the complexities of the data warehouse and simplifies the data mining process. The concept may lead to set trends of storing pre-computed values in data warehouses. This also may be note that the use of subject

oriented computational models also reduces the complexities in data warehouses.

4. To integrate data warehousing, Decision Support System (DSS) and agent technology in subject an oriented data warehouse. There are very few data warehouses are found that have integration with DSS. Further, integrated data mining with agent technology is also focused in this work. However, using Multi Agent System (MAS) for agricultural data mining is a challenging approach. To emphasis on integration of these techniques is another objective of the thesis.

5. To make a comparative review and to contribute in respective and relative works by other researchers.

CHAPTER 2

Literature Review

In this chapter we look through the existing technologies and concepts that are important for conceptualizing our work. We have studied major issues of Indian agriculture. We have also covered the existing technologies used for developing DSS in the field of agriculture. The chapter focuses on the studies of the present Indian agriculture scenario and also covers the findings of DSS that are developed or suggested by previous researchers. The chapter significantly concluded with the study of pre-existing Data Mining Agents that are already used in deferent sectors such as CRM and e-commerce. A brief study about use case of GIS tools in agriculture sector are also discussed in this chapter.

Archaeological and literature evidence have reported that practice of systematic Agriculture in India was began in around 9000 BC. This was the time when many of the old inhabited civilisation started growing and establishing, the north Indian people had earned ability to cultivate plants and food-grains crops such as barley, wheat and jujube. They were well aware about primitive techniques of animal husbandry. Around the 5000 BC they learned to cultivate Cotton, that became a region to form a commercial industry in India. With the help of cotton spinning and fabrication techniques they were soon become well-developed industry for cotton, that further incorporate many others methods to start modern industrialization of India. Around 4500 CE, Indian started systematic irrigation and grow more crops that bring prosperity to this region. As a result, Indian civilization start growing and leading to more planned agriculture system and developed a calendar of systematic cropping. Later, they also started using proper drainage and sewage system for irrigation. They also use to storage water in artificial reservoirs.

Around 200 CE, the people of south India, who were the native of today's Tamilnadu of India start cultivation of wide variety of crops such as sugarcane, rice, black pepper, coconuts and beans. During this time the trade of spices picked the momentum and Indian spices mainly, cinnamon and black pepper were start being shipped to the Mediterranean. The next revolution was "crystallized sugar" that was discovered around 320 CE and soon after application sugar-refining technology it also become a major thing for trade. (History of Organic Farming." WWOOF India. WWOOF India Organics, 2007)

As technologies were being further advanced, irrigation systems were introduced which gave rise to economic growth. The landscape was divided into agricultural 'zones', each producing rice, wheat or millets. Rice production was dominant in Gujarat while wheat dominated north and central India. Other crops to be cultivated were introduced during this period such as tobacco, tea, coffee, pineapples and papaya. From the 18th to the 20th century, India was under British control. The combination of various factors, especially the war in 1918 and being under British rule stagnated India's agricultural economy. Very few of India's commercial crops such as cotton, indigo, opium, and rice made it to the global market. Food and non-food outputs were declining as population was increasing, leaving India in an acute crisis. As the market for irrigation developed, community effort and private investment soared. This eventually helped get India's agriculture back on its feet.

Since India's independence, food and cash crop supply has greatly improved with the establishment of special programs such as The Grow More Food Campaign in the 1940s and the Integrated Production Program in the 1950s. Land reclamation and development, mechanization, electrification, and the use of chemicals soon followed agricultural development. Before India's agricultural and economic fall under British control, India's entire agriculture was practiced organically; materials like fertilizers and pesticides were obtained from plant and animal products. Organic farming shifted to chemical farming in the 1960s when the Green Revolution became the government's most important program for sustaining a rich and stable agricultural economy.

2.1 Indian Cropping System

The Indian climate has diversified parameters, which determine overall agro-ecological setting of Indian agricultural system. The Indian agriculture is mainly based on the two factors; the climate and soil type on which the appropriateness of a particular crop for cultivation is normally based. In this study these two are chosen as the essential ingredients for building agricultural data warehouse. The climatic of north India is a humid subtropical climate. That is very extremely hot in summers and temperature increase up to 50° C, while it is coldest in the winter when the temperature generally falls down up to 0 °C and even low. On the basis of the climate Indian cropping system is categorized in three types of crop seasons namely *Kharif, Rabi* and *Zaid.*

The Rainy season *(Kharif)* started with the arrival of Monsoon in Southwest regions under which the cultivation of tropical crops such as rice, cotton, jute, jowar, bajra and tur are cultivated. The Winter season *(Rabi)* starts with season beginning of winter in October-November and ends in March-April. Commercial Cropping *(Zaid)* is a short duration summer cropping that begins after harvesting of Rabi crops. Some region of India also adopted mixed cropping to cultivate commercial crops. There are four types of cropping systems are practiced in India:

1. Rainy Season Cropping Systems: In this system of cropping, Rice, Sorghum, Pearl Millet (*Bajra*), Maize, Groundnut and Cotton are grown.
2. Winter Cropping Systems: In this system, wheat, barley and oats, sorghum and chickpea are grown.
3. Plantation and other commercial crops: Sugarcane, Tobacco, Potato, Jute, Tea, Coffee, Coconut, Rubber, Spices and condiments are important crops are grown in this system.
4. Mixed Cropping: In this system, pulses and some oilseeds are grown with maize, sorghum and pearl millet.

On the basis of climate compatibility and availability of means of artificial irrigation three types of cropping system are adopted by India farmers:

- **Mono-Cropping or Monoculture:** In this system, only one crop is cultivated on a particular farm land every year. No other crop is grown due to r lack of suitability of soil type of climate conditions.

- **Multiple-Cropping:** In this system, farmers are supposed to cultivate two or more crops on farm land in every calendar year with efficient input management and traditional agricultural practices. It includes inter-cropping, mixed-cropping and sequence cropping.
- **Inter-cropping:** In this system, farmers grow two or more crops simultaneously on the same field in one calendar year. This type of cropping in very common in north India especially for *rabi* crops like wheat.

As India has diversified agro-climatic zone, managing and controlling provided to present time is unfortunately not sufficient good production. Our farming system must be aligned to the modern cropping system and we have to use advance techniques for gain in food grain crop, because due to profit maximization farmers are also inclined towards commercial crops that recorded a noticeable increase in the production of individual crops. The traditional agricultural practices in Indian are still required intensive planning of crop-sequencing, inter-copping and multiple-cropping. The Indian Cropping system can be explained as follows:

2.1.1 The Crops

There are three different types of crops that are cultivated throughout India. Each type is grown in a different season depending on their compatibility with certain climate conditions. *Kharif* crops are grown at the start of the monsoon until the beginning of the winter, relatively from June to November. Examples of such crops are rice, corn, millets, groundnut, *moong*, and *urad*. *Rabi* crops are sown from the beginning of the winter until the beginning of the summer, generally from October to April. Crops grown at this time of year are wheat, barley, grain, oilseeds, and more. *Zaid* crops are rarer than the prior two because they are grown in the short season of the summer. Watermelons and cucumbers are examples of *Zaid* crops. ("Global climate change and Indian agriculture: impacts, adoption and mitigation", Aggarwal, P. K 2008)

2.1.1.1 Cropping Pattern

The cropping pattern is a diversified concept that changes with geographical and seasons that is respectively depending on soil type and climate. It can be defined as sequence of sowing crops in a particular climate and spatial arrangement on a particular. In India, the cropping pattern are determined by rainfall, climate, temperature, soil type and technology. The Farmers choice of cultivating the crops is based on Physical, Social, Economic factors. Many farmers cultivate various crops at their farms and rotate a particular crop combination over a period.

The choice of crops is further depending on influence of several other factors related to agricultural infrastructure, socio-economic factors and technological developments. The infrastructure facilities include means of irrigation, processing, transport, cold storage and marketing of the yield. The Socio-economic factors are land ownership, size and type of land holding, household needs of food, fuel, fibre and finance, labour availability etc. and the technological factors include improved varieties, cultural requirements, mechanization, plant protection, access to information, etc. such advance cropping practices may increase the productivity of crops and could also maintain fertility of the soil.

Table: 2.1.1.1- ***Agro-Climatic Conditions of Major Indian Crops***

Food Grains	Agro-Climatic Condition		
	Temperature	**Rainfall**	**Soil Type**
Rice	22 -32 °C	150-300 cm	Deep clayey and loamy soil
Wheat	10-15 °C (Sowing time) 21-26 °C (Ripening & Harvesting)	75-100 cm	Well-drained fertile loamy and clayey loamy
Millets	27-32 °C	50-100 cm	They are less sensitive to soil deficiencies. They can be grown in inferior alluvial or loamy soil
Grams	20-25 °C (Mild cool & Dry Climate)	40-45 cm	Loamy Soil
Sugar Cane	21-27 °C	75-150 cm	Deep rich loamy soil
Cotton	21-30 °C	50-100 cm	Black soil of Deccan and Malwa Plateau. However, it also grows well in alluvial soils of the Sutluj-Ganga plain and red and laterite soils of the peninsular region
Oilseeds	20-30 °C	50-75 cm	Well drained light sandy loams, red, yellow and black soils are well suited for its cultivation.
Tea	20-30 °C	150-300 cm	Well drained, deep friable loamy soil.
Coffee	15-28 °C	150-250 cm	Well drained, deep friable loamy soil.

Source: Jagran Josh, General-Knowledge-Cropping Patterns and Cropping Systems in India

The cropping patterns in India can also be represented by the major crops of the season or in this study we used the term "crop-set" to explain the intent the main crop and all other possible alternative crops. For example, wheat, barley and oats, are taken as one category. It is also useful to categorize a "crop-set" by its sowing time for example sowing calendar month, and other agro-climatic condition such as required temperature, humidity, rainfall etc. The following table explain major crop-sets, with their required agro-climatic condition.

2.1.1.2 Regional Distribution of Crops

Table: 2.1.1.2- Cropping Regions Across the India

	Crops	States
Cereals	Wheat	Uttar Pradesh, Punjab and Haryana
	Rice	West Bengal, Andhra Pradesh, Chhattisgarh and Tamil Nadu
	Gram	Madhya Pradesh and Tamil Nadu
	Barley	Maharashtra, Uttar Pradesh and Rajasthan
	Pearl Millet (*Bajra*)	Maharashtra, Gujarat and Rajasthan
Cash Crops	Sugarcane	Uttar Pradesh and Maharashtra
	Poppy	Uttar Pradesh and Himachal Pradesh
Oil Seeds	Coconut	Kerala and Tamil Nadu
	Linseed	Madhya Pradesh and Uttar Pradesh
	Groundnut	Andhra Pradesh, Gujarat and Tamil Nadu
	Rape & Mustard	Rajasthan and Uttar Pradesh
	Sesame	Uttar Pradesh and Rajasthan
	Sunflower	Maharashtra and Karnataka
Fibre Crops	Cotton	Maharashtra and Gujarat
	Jute	West Bengal and Bihar
	Silk	Karnataka and Kerala
	Hemp	Madhya Pradesh and Uttar Pradesh
Plantations	Coffee	Karnataka and Kerala
	Rubber	Kerala and Karnataka
	Tea	Assam and Kerala
	Tobacco	Gujarat, Maharashtra and Madhya Pradesh
Spices	Pepper	Kerala, Karnataka and Tamil Nadu
	Cashew Nuts	Kerala, Tamil Nadu and Andhra Pradesh
	Ginger	Kerala and Uttar Pradesh
	Turmeric	Andhra Pradesh & Odisha

Source: Jagran Josh, General-Knowledge-Cropping Patterns and Cropping Systems in India

The soil is one of the most important components for sustainability of cropping system. it is the vitally important natural medium for the growth of vegetation. The soil is defined as, the upper layer of earth in which plants grow. Scientifically, every land type and regions have a particular type of soil consisting of a mixture of various organic element and minerals with clay and rock particles. With the study purpose the Indian soil is categorised in various soil type based on the minerals and organic elements these regions suitable for particular type of crops. The above table is explaining it better.

2.1.1.3 Crop Rotation

Crop rotation can be classified as a type of subsistence farming if there is an individual or communal farmer doing the labour and if the yield is solely for their own consumption. It is characterized by different crops being alternately grown on the same land in a specific order to have more effective control of weeds, pests, diseases, and more economical utilization of soil fertility. In India, leguminous crops are grown alternately with wheat, barley, and mustard. An ideal cropping system should use natural resources efficiently, provide stable and high returns, and avoid environmental damage.

2.1.1.4 Reasons for Crop Rotation

Many farmers in India utilize the crop rotation system to improve or maintain soil fertility, check erosion, reduce the build- up of pests, spread the workload on family labor, mitigate the risk of weather changes, become less reliant on agricultural chemicals, and increase the net profit.

Different Sequences of Crop Rotation

- **Two Crops Rotation**

  i.e.: Year 1: Wheat => Year 2: Barley => Year 3: Wheat again

- **Three Crop Rotation**

  i.e.: Year 1: Wheat => Year 2: Barley => Year 3: Mustard => Year 4: Wheat again.

(Ramesh, P., N. R. Panwar, A. B. Sing, S. Ramana, Sushil Kumar Yadav, Rahul Shrivastava, and A. Subba Rao. "Status of Organic Farming in India.")

2.1.2 The Soil

In ancient times, soils were classified into only two groups - *Urvara* and *Usara*, which literal meaning were fertile and sterile, respectively. In medieval times, the soils were classified on the basis of the physical features such as texture, colour, slope of land and moisture content in the soil. And in that time the soils were identified as sandy, clayey, silty and loamy, etc. In few regions they also classify the soil on the basis of colour such as red soil, yellow soil, etc.

Table: 2.1.2- Types of Soils in India

Soil Type	General Characters / Position in entire world soil classification	Area	% of Total area of India
Inceptisols	These are usually the weakly developed young soil though they are more developed than entisols.	130372.9	39.74
Entisols	Usually young or underdeveloped. Lack vertical development of horizons. These are less fertile soils.	92131.71	28.08
Alfisols	Pale, grayish brown to reddish in colour with moderate-to-high reserves of basic cations and are fertile. However, their productivity depends on moisture and temperature. They are supplemented by the moderate application of lime and other chemical fertilizers.	44448.68	13.55
Vertisols	These are expandable clay soils, composed of more than 30 per cent clays. Vertisol clays are black when wet and become iron hard when dry. When drying, Vertisols crack and the cracks widen and deepen as the soil dries; this produces cracks 2-3 cm wide. These are productive soils. The regur soils of India are an example of vertisols.	27960	8.52
Aridisols	Aridisols have a very low concentration of organic matter, reflecting the paucity of vegetative production on these dry soils. Water deficiency is the major defining characteristic of Aridisols. Also required is sufficient age to exhibit subsoil weathering and development	14069	4.28

Ultisols	Commonly known as **red clay soils,** they are defined as mineral soils which contain no calcareous (calcium carbonate containing) material anywhere within the soil, have less than 10% weatherable minerals in the extreme top layer of soil, and have less than 35% base saturation throughout the soil.	8250	2.51
Mollisols	Mollisols have deep, high organic matter, nutrient-enriched surface soil typically between 60 and 80 cm in depth.	1320	0.4
Others	Other land area includes Snowfields, mountains or desert area. That may not fit for agriculture purpose.	9503.1	2.92
		Total =	100

Source: Soils of India, National Bureau of Soil Survey and Land Use Planning, Publication Number 94

The Indian council of Agricultural Research (ICAR) set up an All India Soil Survey Committee in 1953. On the basis of Soil Taxonomy defined by United States Department of Agriculture (USDA), the ICAR has divided Indian soils into eight major groups. The characteristics and proportion from total area is given in in the above table. The geographical classification done by ICAR is also given as follows:

Figure: 2.1.2- Types of Soils in India

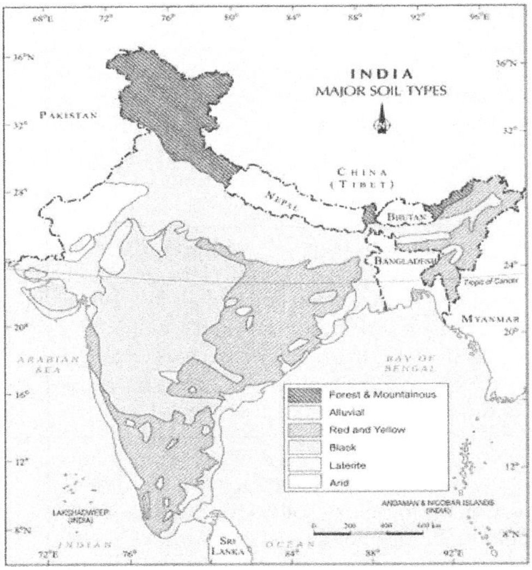

Source: Soils of India, National Bureau of Soil Survey and Land Use Planning

On the basis of genesis, colour, composition and location, the soils of India are classified into eight groups, in this study they are named as Soil Type 1, Soil Type 2 and so on.

1. Alluvial soils
2. Black soils
3. Red and Yellow soils
4. Laterite soils
5. Arid soils
6. Saline soils
7. Peaty soils
8. Forest soils

2.1.3 Fertilizers

The fertilizer industry in India has grown tremendously in the last 30 years. The Government is keen to see that fertilizer reaches the farmers in the remote and hilly areas. It has been decided to decontrol the prices, distribution and movement of phosphatic and potassic fertilizers. Steps have been taken to ensure an increase in the supply of non-chemical fertilizers at reasonable prices. There are 53 fertilizer quality control laboratories in the country. Since bio--fertilizers are regarded as an effective, cheap and renewable supplement to chemical fertilizers, the Government is implementing a National Project on Development and Use of Bio-fertilizers. Under this scheme, one national and six regional centres for organizing training, demonstrating programs and quality testing of has been taken up.

2.1.4 Irrigation

Irrigation in India refers to the supply of water from Indian rivers, tanks, wells, canals and other artificial projects for the purpose of cultivation and agricultural activities. In country such as India, 64% of cultivated land is dependent on monsoons. The economic significance of irrigation in India is namely, to reduce over dependence on monsoons, advanced agricultural productivity, bringing more land under cultivation, reducing instability in output levels, creation of job opportunities, electricity and transport facilities, control of floods and prevention of droughts.

2.1.4.1 Geography of Irrigation in India

Irrigation farming is very important for crop cultivation in regions of seasonal or low rainfall. Western U.P., Punjab, Haryana, parts of Bihar, Orissa, A.P., Tamil Nadu, Karnataka, and other regions thrive on irrigation and generally practice multiple or double cropping. With irrigation, a large variety of crops can be produced such as rice, sugarcane, wheat, and tobacco. (Bhaduri, Anik, Upali Amarasinghe, and Tushaar Shah. "Future of Irrigation in India").

2.1.5 Technological Needs for Agriculture Development

It is apparent that the tasks of meeting the consumption needs of the projected population are going to be more difficult given the higher productivity base than in 1960s. There is also a growing realization that previous strategies of generating and promoting technologies have contributed to serious and widespread problems of environmental and natural resource degradation. This implies that in future the technologies that are developed and promoted must result not only in increased productivity level but also ensure that the quality of natural resource base is preserved and enhanced. In short, they lead to sustainable improvements in agricultural production. Productivity gains during the 'Green Revolution' era were largely confined to relatively well-endowed areas. Given the wide range of agroecological setting and producers, Indian agriculture is faced with a great diversity of needs, opportunities and prospects. Future growth needs to be more rapid, more widely distributed and better targeted. Responding to these challenges will call for more efficient and sustainable use of increasingly scarce land water and germplasm resources. Technical solutions required to solve problems will be increasingly location-specific and matched to the huge agro-ecological/climatic diversity. Detailed indigenous knowledge and greater skills in blending modern and traditional technologies to enhance productive efficiency will be more than ever before, key to the farming success and sectoral growth. Most technological solutions will have to be generated and adapted locally to make them compatible with socio-economic conditions of farming community.

New technologies are needed to push the yield frontiers further, utilize inputs more efficiently and diversify to more sustainable and higher value cropping patterns. These are all knowledge intensive technologies that require both a strong research and extension system and skilled farmers but also a reinvigorated interface where the emphasis is on mutual exchange of information bringing advantages to all. At the same time potential of less favored areas must be better exploited to meet the targets of growth and poverty alleviation. These challenges have profound implications for products of agricultural research. The way they are transferred to the farmers and indeed the way research is organized and conducted. One thing is, however, clear – the new generation of technologies will have to be much more site specific, based on high quality science and a heightened opportunity for end user participation in the identification of targets. These must be not only aimed at increasing farmers' technical knowledge and understanding of science-based agriculture but also taking advantage of opportunities for full integration with indigenous knowledge. It will also need to take on the challenges of incorporating the socio-economic context and role of markets. With the passage of time and accelerated by macro-economic reforms undertaken in recent years, the Institutional arrangements as well as the mode of functions of bodies responsible for providing technical underpinning to agricultural growth are proving increasingly inadequate. Changes are needed urgently to respond to new demands for agricultural technologies from several directions. Increasing pressure to maintain and enhance the integrity of degrading natural resources, changes in demands and opportunities arising from economic liberalization, unprecedented opportunities arising from advances in biotechnology, information revolution and most importantly the need and urgency to reach the poor and disadvantaged who have been by passed by the green revolution technologies.

Another important implication of increasing globalization relates to the need for greater attention to the quality of produce and products both for the domestic and the foreign markets. This would imply that production must be tuned to actual rapidly changing product demand. Such adaptation to global markets would require state of the art research, which can be achieved only by setting global standards of research, focus on well-defined priorities and mechanisms which permit close interaction of farmers with researchers, the private sector and markets.

2.1.6 Agricultural Research and Information Technology

The apex body for education, research and extension education in the field of agriculture is the Indian Council of Agricultural

Research (ICAR), established in 1929. India's transformation from a food deficit to a food surplus country is largely due to ICAR's smooth and rapid transfer of farm technology from the laboratory to the land. ICAR discharges its responsibilities through 43 research institutes, four national research bureaus, 20 national research centres, nine project directorates, 70 all-India coordinated research projects, and 109 *Krishi Vigyan Kendras* (farm science centres). Besides, the program of Agricultural Education is coordinated by ICAR with the curricula and other normative guidance given to the 26-Agricultural Universities and four National Research Institutes.

The Integrated national agricultural resources information system (INARIS) is a state of art Central Data Warehouse (CDW) of agricultural resources of the country has been developed INARIS. A work "Improving Agricultural Knowledge and Innovation Systems" is also published in OECD Conference Proceedings. Reader also may find the details in article "Information Systems and Software Development at IASRI: An Overview" (P.K. Malhotra and R.C. Goyal).

2.2 MIS and DSS for Agriculture

Data collected in agricultural research in laboratory and on-farm research are subjected to descriptive, inferential, diagnostic and analytical processing. The statistical packages cover most of the statistical procedures often required for data analysis, outstanding graphics capabilities, data handling tools, and are user friendly.

2.2.1 Information Systems for Agriculture

With the installation of B-4700 in 1977, the activity for development of information systems and data base applications was initiated. A group of scientists designed and implemented a data base comprising of the bio-data of agricultural research scientists under the ICAR institutes. A selective dissemination information service based on AGRIS data base available on tapes received from Vienna every month was in operation since 1978. This enabled the scientists in India to obtain references on selected topics of their interest in the field of agricultural science and technology. During 1980-84 necessary software were developed for the information storage and retrieval system for National Index of Field Experiments and National Index of Animal Experiments. During 1990-2000 a number of information systems were developed, to name a few National Information System on Agriculture Education (NISAGE); Project Information and Management System (PIMS); Agriculture Research Data Information System - ARDIS. (Information Systems and Software Development at IASRI: An Overview P.K. Malhotra and R.C. Goyal.)

2.2.1.1 National Information System on Agricultural Education Network in India (NISAGENET)

On the recommendations of the National Statistical Commission (NSC-2001) and the Department of Secondary & Higher Education, Ministry of Human Resource Development, Government of India so as to maintain and update the data regularly on parameters related to agricultural education in India. (Goyal et al., 2006, 2007). NISAGENET is currently coordinating with 42 agricultural organizations in the country for desired data collection on agricultural education in India. These include State Agricultural Universities, ICAR Deemed Universities, and Central Universities in agricultural education and other private colleges/ centres affiliated to these organizations. The database contains the information on various aspects such as Academic data of the universities, Infrastructural facilities, Budget provision, Manpower employed and R&D activities. This on-line system provides an integrated platform for collection, analysis and dissemination of information on programs and activities of the agricultural education system in India through internet service mode.

2.2.1.2 Project Information Management System Network (PIMSNET)

PIMSNET was designed, developed and implemented under the World Bank funded National. The Agricultural Technology Project's (NATP) sub-project 'Institutionalization of Research Priority Setting, Monitoring and Evaluation and Networking of Social Scientists' under Project Monitoring and Evaluation (PME) Component of Organisational & Management (Goyal et. al., 2002, Sharma et al., 2005). PIMSNET was used for Monitoring and Concurrent Evaluation (M & CE) of 845 research projects under NATP so as to effectively assess research efforts against well-defined targets, avoid duplication of research efforts and provide feedback to research planning process It had four modules viz., Data Management, Monitoring, Reports and Query and Administration.

2.2.1.3 Online Information System on Personal Management in ICAR

Personnel Management Information System Network (PERMIS net) for ICAR has been developed and implemented for providing information for systematic planning and development of human resources in ICAR (Balbir et al., 2004, 2005, 2008), (Shashi et al., 2004), (Alka et al., 2005), (Farooqi et al., 2005). The system provides information on scientists, technical personnel, administrative and supporting staff on personal and professional attributes of each employee, besides information on cadre strength and institutional structure to help the management in taking appropriate timely decisions.

2.2.2 DSS for Agriculture

Complementing the development of computer systems, the last few decades have seen significant progress in the development and application of decision theory. Primarily as an outgrowth of the field of operations research, decision theory has become much easier to apply to very complex decisions. One of the key advances has been the development of approaches to making decisions with multiple objectives. Keeney and Raiffa (1993), March (1994), and Hammond and Keeney (1999) have written good introductory books on decision theory with application to multiple objective decision-making. Beinat and Nijkamp (1998) present a number of applications of multi-objective theory to land management. Other examples of multiple objective decision-making in natural resources include the application of Saaty's (1990) Analytic Hierarchy Process, or AHP, to natural resources (Schmoldt et al., 2001) and for watershed management (de Steiguer et al., 2003), and DEFINITE by Janssen and van Herwijnen (1994). Since the early 1990s, many natural resource applications of DSS have been developed.

2.2.2.1 Decision Support Systems (DSS) in Crop Protection

The Central Institution for Decision Support Systems (DSS) in Crop Protection (German acronym ZEPP) is to develop, collect and examine existing forecasting and simulation models for important agricultural and horticultural pests and diseases and to adapt these models for practical use. Mr Racca Paolo and five other co-researchers presented a paper explain the work in an article "Decision Support Systems in Agriculture: Administration of Meteorological Data, Use of Geographic

Information Systems (GIS)and Validation Methods in Crop Protection Warning Service". (Racca Paolo, Kleinhenz Benno, Zeuner Thorsten, Keil Barbara, Tschöpe Beate and Jung Jeanette "Decision Support Systems in Agriculture: Administration of Meteorological Data, Use of Geographic Information Systems (GIS) and Validation Methods in Crop Protection Warning Service"). In this work many weather-based forecasting models for pests and diseases have been successfully developed. The occurrence of diseases/pests and periods of high-intensity attacks can be calculated with high accuracy.

2.2.2.2 DSS in Agriculture, Food and the Environment

A similar work we have received from Basil Manos, Konstantinos paparazzo, Nikolaos Matsatsinis, Jason Papathanasiou, June 2010, from the various university of Greece. The researchers developed "Decision Support Systems in Agriculture, Food and the Environment: Trends, Applications and Advances" presents the development of DSS for managing agricultural and environmental systems, focusing on the exposition of innovative methodologies, from web-mobile systems to artificial intelligence and knowledge-based DSS, as well as their applications in every aspect from harvest planning to international food production and land management.

2.2.2.3 Decision Support System for Agriculture Using NLP

Prof. Mrs. J.R.Prasad, Prof. R.S.Prasad, Dr. U.V.Kulkarni suggests development of a decision support system for agriculture based on the natural language processing. In this article he has quoted "The analytical data about the rainfall pattern, soil structure of the area will be maintained at back end, the system will retrieve the information based on the interaction with the user, which will be a farmer in this case".

The authors aim to provide a user-friendly decision support system. His work is published in proceedings of the International Multi Conference. However, many DSS has been found in the literature review but we still find the DSS for estimate crop production and to answers for farmer queries is still a need.

2.3 Data Mining in Agriculture

It consists in the application of data mining techniques to agriculture. Recent technologies are nowadays able to provide a lot of information on agricultural-related activities, which can then be analyzed in order to find important information. (Mucherino, A.; Papajorgji, P.J.; Pardalos, P. 2009). A related, but not equivalent term is precision agriculture. There are many conferences organized every year on data mining techniques and applications, but rather few of them consider problems arising in the agricultural field.

2.3.1 Data Mining Agent for Agriculture

A data mining agent is a pseudo-intelligent computer program designed to ferret out specific types of data, along with identifying patterns among those data types. Data mining agents are typically used to detect trends in data, alerting organizations to paradigm shifts so effective strategies can be implemented to either take advantage of or minimize the damage from alterations in trends. In addition to reading patterns, data mining agents can also "pull" or "retrieve" relevant data from databases, alerting end-users to the presence of selected information. Conceptualize a data mining agent as a very limited type of virtual employee. In effect, a data mining agent is nothing more than an employee tasked with sorting through employee records to perform one or more very specific jobs. For example, a data mining agent could be programmed to monitor stock prices for a specific range of companies, throwing up a red flag if it notices any substantial deviations from historical trends. Data mining agents are a bit like a smoke alarm; they only throw up signals if something is actually happening in the system. In recent years, an increasingly remarkable trend in both areas is that integrating agents into data mining systems in order to extract knowledge from the distributed and heterogeneous data. These agents use different methods to handle different types or parts of information in heterogeneous data sets. The results are combined to get integrated results in this model that proposes three types of agents, firstly is manager agent which create, suspect, resume miner agent and cooperator agent, secondly is miner agent, which apply multiple algorithms to data sets for extract knowledge, finally is cooperator agent, which is to integrated partial and local knowledge to more high-level knowledge.

2.3.1.1 Weather Information Agent

The Weather Information Agent aims to support the decision-making process regarding the issues that concern weather conditions. This Agent gathers information from meteorological services, combining it with measurements from the cultivation site. Based on this information it provides short term prediction of weather conditions. Moreover, the Weather Information Agent cooperates with the historic module of the E-Learning Agent in a twofold manner: first it exploits information from the historic module to support the weather prediction process; secondly, the information gathered by the Weather Information Agent is used to enrich the historic module.

2.3.1.2 E-Commerce Agent

The E-Commerce Agent aims to support the sales through internet of agricultural products and it can also support the farmer in buying row material through internet. The E-Commerce Agents comprises an intelligent unit to use the market information stored in the historic module, in order to decide whether each buying or selling action is profitable for the farmer. The Logistics Agent aims to support the farmer concerning decisions on planning and controlling an efficient flow and storage of raw materials, intermediate and final products from point of origin to point of consumption, so as to achieve specific performance requirements, while simultaneously satisfying constraints. (Berlo van, 1993; Folinas et al.,2003; Biere, 2001). The performance requirements concern the product cost and quality, the satisfaction of the market demand, the optimal exploitation of the processing equipment and cultivation area, etc. The restrictions concern the storage capacity, the maximum allowed time of storage, the available cultivation capacity and the processing capacity, etc. The planning provided by the Logistics Agent should take into account time changing and uncertain factors, like the weather conditions and the demand of the market. The Logistics Agent may consider the whole logistical chain incorporating agriculture, processing industry and market, as was proposed in (Berlo van, 1993); in this case, the DSS should communicate with corresponding information systems of the processing industry and market sectors, in order to derive the data required.

2.3.2 Multi-Agent Simulation Models in Agriculture

Intelligent agents are used in various complex systems like biology, artificial intelligence (AI), computer networking, robotic systems, computer games, and military defence systems, transportation logistics, GIS and many other fields. Precision agriculture complex systems which are based on multi-agent technology can be divided in two different multi-agents-based technology types: homogeneous robot multi-agents system (RMAS) (Quinn M 2003.) (Candea C., Hu H., Iocchi L., Nardi D., Piaggio M., 2001) and heterogeneous software multi-agents system (SMAS). Each subsystem has communication technology between agents, deliberation mechanism, data acquiring from influencing environment and other collaborative agents, decision making mechanisms, learning technologies, goal - based behaviour ability. Multi-Agent Simulation (MAS) models are intended to capture emergent properties of complex systems that are not amenable to equilibrium analysis. They are beginning to see some use for analysing agricultural systems. The paper reports on work in progress to create a MAS for specific sectors in New Zealand agriculture.

2.4 GIS and Remote Sensing Applications

An Agricultural decision support system includes automated risk assessment analysis. This would allow the farmer to choose a specific cropping system for their farm location. The system would calculate the statistical probabilities of success by including climatic variation, soil properties, and crop species requirements. Another addition will be the ability to perform cropping system analysis and risk assessment at field the level.

2.4.1 Application of GIS in Agriculture

Satellite based remote sensing technique combined with limited field survey provides valuable information in quick time. The data generated from satellite imageries provide information on the status of crops under observation at any particular date, and can be processed quickly. This inherent advantage provides a synoptic view, making it possible to analyse the status and trend of transplantation, crop growth and harvesting throughout the study area. Wide spread application of remote sensing and GIS will be helpful for creating a systematic and sharable database on crop related issues and answer the unanswered questions of different stakeholders. The figure below shows a general methodology adopted for extraction of data using remote sensing & GIS and its dissemination to different stakeholders.

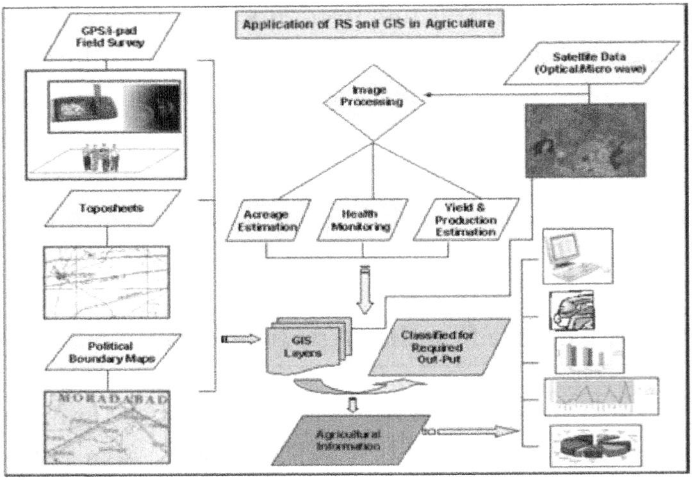

Figure : 2.4.1 (a)- Application of GIS in Agriculture

Figure :2.4.1 (b)- Satellite based estimation of Basmati acreage with different GIS layers

The information sought by the exporters includes variety wise acreage estimation, production estimation, health monitoring of crops, and knowledge on draught and flood conditions. The above Map -*Fig: 2.4.1 (b)* is showing distribution of *Basmati* in the state of Haryana during the *Kharif* 2005.

2.4.2 Remote Sensors for Precision Farming

Precision farming needs information about mean characteristics of small, relatively homogeneous management zones. This is for Data acquisition of the farms to find the soil, vegetation and other parameters that are amenable for remote sensing. Remote sensing techniques play an important role in precision farming by providing continuous acquired data of agricultural crops. Remote sensors image vegetation, which is growing on different soil types with different water availability, substrate, impact of cultivation, and relief Sensors use for the following applications:

- Soil Properties Sensing: Soil Texture, Structure, and Physical Condition Soil Moisture; Soil Nutrients.
- Crop Sensing: Plant Population; Crop Stress and Nutrient Status.
- Yield Monitoring Systems: Crop Yield; Harvest Swath Width; Crop Moisture:

2.4.3 GIS and Land Use

Geographic Information System applied to an agricultural landscape can effectively capture, store, analyse and display information that is geographically based. Of its many benefits, GIS can improve our understanding of farming areas, help promote agricultural development and assist in identifying and handling issues important to strengthening farming. The

Ministry of Agriculture and Lands has been working with many local governments in undertaking land use inventory projects in their farming areas. A variety of maps and statistics can be created from the land use inventory, including primary land use activities, agricultural land use changes, and activities along the urban-rural edge. The Ministry has developed a series of tools to assist local governments that are interested in conducting a land use inventory. These tools include an Access database and coding system for recording information and a 40-page guide on the land use inventory methodology.

The method developed by the Ministry involves a team of two surveyors conducting a "windshield" survey. Using a combination of drive-by observations and aerial photographic interpretation, the survey team examines each legal parcel within the study area, recording both the land covers (e.g., buildings, crops) and land use activities (e.g., agricultural, residential). The information for each parcel is entered into a database and then linked to a GIS layer of the legal parcels. A GIS user can then query and map the land use inventory information. Alternative cropping system research currently being conducted at these sites are demonstrating substantial increases in annual yields due to improved soil quality and available water.

2.4.3.1 Agri-FACTs

The Agri-FACTs program was created using ArcView 3.1, ArcView Avenue, ArcView Spatial Analyst extension, and the ArcView Internet Map Server (IMS) extension with an HTML graphical user interface. We chose the ArcView IMS because the system deals with large raster datasets. Map Objects IMS does not have the GRID analysis capability that was required for the system. Using ArcView 3.1 with the Spatial Analyst and ArcView IMS extensions allowed Agri-FACTs to operate very quickly and accurately. Original surface models were transformed from a GRASS GIS format into an Arc/INFO GRID format and re projected using Arc/INFO 7.1. Agri-FACTs also exists as a standalone ArcView project. (Vernon L. Thomas, Robert M. Aiken, William Waltman).

Agri-FACTs is designed to be as user friendly as possible. An accessible and easy-to-use format may make Agri-FACTS a decision support system of choice for producers across the region who are seeking useful information on alternative cropping systems. The process simply involves zooming and panning to the farm location, clicking on the Agri-FACTs button, and finally clicking on the general point location of their farm. The system then automatically collects the cell value from each surface model and creates an HTML page report that the user can then analyse and/or print.

The Agri-FACTs HTML report displays the values for all six models (PET, AWB, SRPG, PRECIP, RZWHC, and GDD) for the selected farm point. The report allows the farmer to individually compare each of the soil and climatic values found at the farm location to the soil and climatic values found at each of the five research sites. The farmer can then determine which research station(s) has the most similar soil and climatic characteristics to their farm. The report contains hyperlinks to each of the experiment stations allowing the farmer to go directly to the research site's web pages and examine the cropping system research results. It is at this level that Agri-FACTs facilitates decision support regarding new cropping systems that the producer may elect to consider for his farm to replace the wheat-fallow system most commonly used. The farmer can then take this information to his extension agent or district conservationist for a more explicit analysis.

2.5 Chapter Conclusion

In this chapter we have made review of various concepts, tools and systems used for agriculture data warehouse. We have described the literature of Indian cropping system that includes traditional system and advance technology used in the farming. Various MIS and DSS are also discussed that are implemented or proposed for researchers. Data mining techniques and data mining agents that are used for different branches and applied for DSSs are also reviewed by us. From computational point of view, we have described agent characteristics, multi-agent system paradigm. GIS based applications that are used in agriculture are also reviewed. Some already used GIS applications in the field of agriculture are also discussed. The purpose of the literature reviews it to find the status of work already done and to find the scope of applying these technologies, tools and practices for developing a data warehouse for agriculture. Some observations have been made from the literature review in related area of research:

- It is observed that a large number of Data Warehouses are developed in the field like CRM, Health and others. However very limited numbers of works are available which try to implement the data warehouse technology in the field of agriculture.
- There is no deterministic computational model for estimating the performance of crop.
- Due to complexity of agricultural data keeping and retrieval of agricultural data from data warehouse is a key challenge for researches.
- Storing pre-calculated values in agricultural data warehouse is a key idea for this work.

- I also observed that the use of software agent for data mining purpose in agricultural data warehouse is also lead to useful approach because the software agent can be highly beneficial to achieve high degree of automation.
- The agricultural data is spatial in nature, so it is better to visualize the agricultural data using spatial technology like Geographic Information System. I also include GIS in my work as a useful tool for data visualization.

Keeping the above facts in mind, I have tried to integrate all said technologies of computer science to develop an Integrated Data Warehouse for Indian Agriculture.

CHAPTER 3

Basic Concepts

This chapter includes various theories, practices and common knowledge that is used in this research. The definitions, salient features, functions, procedure and methodology of various concepts for building a data warehouse are discussed in details. The chapter also covers data warehouse terminology and designing process in the first section followed by the various data mining techniques such as neural network and decision tree. Further the chapter also describes the concept of agent technology and multi agent system, and finally includes a brief discussion about visualization geographical information system.

3.1 Data Warehousing

The term *data warehouse* was first used by William Inmon in early 1980's. He defined data warehouse to be a set of data that support DSS and is "subject-oriented, integrated, time-variant, and non-volatile" [Inm95]. A common way of introducing data warehousing is to refer to the characteristics of data warehouse as set by William Inmon.

Subject Oriented: Data warehouses are designed to help you to analyse data related to particular subject. For example, to learn more about your health care data, you can build a warehouse that concentrates on medical issues. The ability to define any data warehouse by its subject matter makes a data warehouse subject oriented.

Integrated: Integration is closely related to subject orientation. Data warehouses must put data from disparate sources into a consistent format. They must resolve such problems as naming conflicts and inconsistencies among units of measure. When they achieve this, they are said to be integrated.

Non-volatile: Non-volatile means that, once entered into the warehouse, data should not change. This is logical because the purpose of a warehouse is to enable you to analyse what has occurred.

Time Variant: In order to discover trends in business, analysts need large amounts of data. This is very much in contrast to online transaction processing (OLTP) systems, where performance requirements demand that historical data be moved to an archive. A data warehouse's focus on change over time is what is meant by the term time variant.

The data warehouse is a relational database that is designed for query and analysis rather than for transaction processing. It usually contains historical data derived from transaction data, but it can include data from other sources, generally data is collected from various related fields. It separates analysis workload from transaction workload and enables an organization to consolidate data from several sources. In every industry, data sets are organized around individual applications to support particular operational systems. These individual data sets have to provide data for specific application to perform specific function efficiently. Therefore, the data sets for each application need to be organized around that specific application. In addition to a relational database, a data warehouse environment includes an extraction, transportation, transformation, and loading (ETL) solution. An online analytical processing (OLAP) engine, client analysis tools, and other applications that manage the process of gathering data and delivering it to the users.

Data Mart: A data mart is the access layer of the data warehouse environment that is used to get data out to the users. The data mart is a subset of the data warehouse that is usually oriented to a specific purpose or theme. In most of the applications each department or functional unit is considered as the owner of its data-mart including all the hardware, software and data. This enables each department to use, manipulate and develop their data any way they see fit, without altering information inside other's data marts or the data warehouse. In order to ensure effective management, development and **deployment** of resource in agriculture sector, we can create separate data-marts for various purpose and functional units. This approach of segregation is useful for intelligent agriculture. As at present the business intelligence applications are used to gather, store, access and analyse heterogeneous data-marts collected through various sources and technologies. The data mart is also useful for routine functions of any particular industry. Ralph Kimball quotes "The data warehouse is nothing more than the union of all the data marts". The data mart is hence a collection of data sets from subject and related areas that are heterogeneous in nature and organized for providing decision support to the business department or industry.

Meta Data: Metadata is often defined as data about data. (Cabinet Office (2006), e-Government Metadata Standard

Version 3.1, e-Government Unit, London). It is "structured information that describes, explains, locates, or otherwise makes it easier to retrieve, use or manage an information resource", especially in a distributed network environment like for example the internet or an organization. Metadata describes how and when and by whom a particular set of data was collected, and how the data is formatted. Metadata is essential for understanding information stored in data warehouses. Metadata is used to provide documentation for data products. A good example of metadata is the cataloguing system found in libraries, which records for example the author, title, subject, and location on the shelf of a resource. Metadata is usually categorized in three types:

- Descriptive metadata describes an information resource for identification and retrieval through elements such as title, author, and abstract.
- Structural metadata documents relationships within and among objects through elements such as links to other components (e.g., how pages are put together to form chapters).
- Administrative metadata helps to manage information resources through elements such as version number, archiving date, and other technical information for purposes of file management, rights management and preservation.

3.1.2 The Need of Data Warehousing

Over the last two decades, a number of experiences and technologies incorporated together to evolve the new field of Data warehousing. It could be understood as a well-organized and resourceful method of managing & reporting data collected from non-uniform and scattered sourced throughout an organization. Due to hundreds of gigabytes of transactions, it is necessary for a data warehouse to be vast conceptually and large storage capacity. Therefore, 'data marts' are often designed for individual department or a product line. A data warehouse system is an influential and necessary platform for merging data from old and new applications. The Data warehouse is an environment, not a product.

It is an architectural construct of information systems that provides users with current and historical decision support information that is hard to access or present in traditional operational data stores. In fact, the data warehouse is a cornerstone of the organization's ability to do effective information processing, which among other things, can enable and share the discovery and exploration of important business trends and dependencies that otherwise would have gone unnoticed. In principle, the data warehouse can meet informational needs of knowledge workers and can provide strategic business opportunities by allowing customers and vendors access to corporate data while maintaining necessary security measures. Here are some of the benefits of a data warehouse:

- With data warehousing, we can provide a common data model for different interest areas regardless of data's source. In this way, it becomes easier to report and analyse information. This feature is useful for managing heterogeneity of agriculture data, as data is collected from various sources and formats.
- Many inconsistencies are identified and resolved before loading of information in data warehousing. This makes

the reporting and analysing process simpler. This is also an advantage for this work as the generated report is simpler it is easier to understand by farmers.

- The best part of data warehousing is that the information is under the control of users, so that in case the system gets purged over time, information can be easily and safely stored for longer time period

3.1.3 Data Warehouse Architecture

Different data warehousing systems have different structures. Some may have an ODS (operational data store), while some may have multiple data marts. Some may have a small number of data sources, while some may have dozens of data sources. In view of this, it is far more reasonable to present the different layers of data warehouse architecture rather than discussing the specifics of any one system. In general, all data warehouse systems have the following layers:

1. **Data Source Layer**
2. **Data Extraction Layer**
3. **Staging Area**
4. **ETL Layer**
5. **Data Storage Layer**
6. **Data Logic Layer**
7. **Data Presentation Layer**
8. **Metadata Layer**
9. **System Operations Layer**

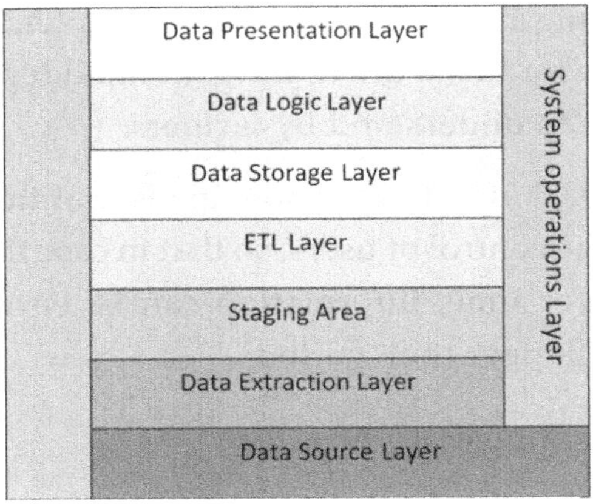

Figure: 3.1.3- Data Warehouse Architecture

This represents the different data sources that feed data into the data warehouse. The data source can be ranges from various and various formats such as plain text files, relational database, other types of databases.

Data Extraction Layer:

Data gets pulled from the data source into the data warehouse system. There is likely some minimal data cleansing, but there is unlikely any major data transformation.

Staging Area:

This is where data sits prior to being scrubbed and transformed into a data warehouse/data mart. Having one common area makes it easier for subsequent data processing or integration.

ETL Layer:

This is where data gains its "intelligence", as logic is applied to transform the data from a transactional nature to an analytical nature. This layer is also where data cleansing happens. The ETL design phase is often the most time-consuming phase in a data warehousing project, and an ETL tool is often used in this layer.

Data Storage Layer:

This is where the transformed and cleansed data sit. Based on scope and functionality, 3 types of entities can be found here: data warehouse, data mart, and operational data store

Data Logic Layer:

This is where business rules are stored. Business rules stored here do not affect the underlying data transformation rules, but do affect what the report looks like.

Data Presentation Layer:

This refers to the information that reaches the users. This can be in a form of a tabular or graphical report in a browser, an emailed report that gets automatically generated and sent every day, or an alert that warns users of exceptions, among others. Usually an OLAP tool and/or a reporting tool is used in this layer.

Metadata Layer:

This is where information about the data stored in the data warehouse system is stored. A logical data model would be an example of something that's in the metadata layer.

System Operations Layer:

This layer includes information on how the data warehouse system operates, such as ETL job status, system performance, and user access history.

3.1.4 Business Wide Data Warehouse Architecture

A business-wide data warehouse is considered one that will support all, or a large part, of a business that requires a more fully integrated data warehouse, with a high degree of data access and usage across departments or lines of business. That is, the warehouse is designed and constructed based on the needs of a business as a whole. It could be considered to be a common repository for decision-support data that is available across the entire organization, or a large subset of that data. The term "business-wide" is used here to reflect the broad scope of data accessing and usages, not to the physical structure of the data warehouse.

3.1.5 Data Warehouse Implementation Approaches

The choice of an implementation approach is influenced by such factors as the current IT infrastructure, the resources available, the architecture selected, and the scope of the implementation, the need for more business-wide data access across the organization, the return-on-investment requirements, and the speed of implementation. The selection of an implementation approach is a very important part of data warehouse design; the decision needs to be made at an early stage of the data warehouse project.

Top-down Implementation: The top-down approach is to implement the data warehouse in a single project phase. A top-down implementation requires more planning and design work to be completed at the beginning of the project. This brings with it the need to involve people from each of the workgroups, departments, or lines of business that will be participating in the data warehouse implementation. Decisions about data sources to be used, security, data structure, data quality, data standards, and an overall data model will typically need to be completed before actual implementation begins.

Bottom-up Implementation: A bottom-up implementation involves the planning and design of a data warehouse without waiting for a more business-wide data warehouse design to be put in place. This does not mean that a more business-wide data warehouse design will not be developed; it will be built incrementally as initial data warehouse implementations expand. This approach is more widely accepted today than the top-down approach, because immediate results from the data warehouse can be realized and used as justification for expanding to a more business-wide implementation. Each implementation approach has advantages and disadvantages. In many cases the best approach may be a combination of the two. One of the keys to this approach is to determine the degree of planning and design that is required for the business-wide architecture to support integration as the data warehouse is being built with the bottom-up approach.

3.1.6 Dimensional Modelling

Dimensional modelling is a set of techniques and concepts used in data warehouse design. It is considered to be different from entity-relationship modelling. Dimensional Modelling does not necessarily involve a relational database. The same modelling approach, at the logical level, can be used for any physical form, such as multidimensional database or even flat files. According to (Kimball 1997).

> *"Data Modelling is a design technique for databases intended to support end-user queries in a data warehouse. It is oriented around understandability and performance. According to him, although transaction-oriented ER is very useful for the transaction capture, it should be avoided for end-user delivery".*
>
> *- Ralph Kimball*

Dimensional modelling always uses the concepts of facts, and dimensions. Facts are typically numeric values that can be aggregated, and dimensions are groups of hierarchies and descriptors that define the facts. The key point on the design of dimensional models is to resolve questions in the format "measures by dimensions." In addition, dimensional models are commonly referred to as star schema as they comprised of a central fact table surrounded by several dimension tables. Within an enterprise data warehouse or data marts, data is fundamentally static, non-volatile and does not get updated. Rather data is inserted or loaded in bulk into the tables in the model utilizing using batch programs or extraction, transformation, & loading (ETL) routines. End-users of dimensional models develop queries that either read or select data, and there is no end-user inserting, updating, or deleting of data. **Some major benefits of the dimensional modelling are**:

1. **Understandability:** Compared to the normalized model, the dimensional model is easier to understand and more intuitive. In dimensional models, information is grouped into coherent business categories or dimensions, making it easier to read and interpret. Simplicity also allows software to navigate databases efficiently. In normalized models, data is divided into many discrete entities and even a simple business process might result in dozens of tables joined together in a complex way.

2. **Query performance:** Dimensional models are more de-normalized and optimized for data querying, while normalized models seek to eliminate data redundancies and are optimized for transaction loading and updating. The predictable framework of a dimensional model allows the database to make strong assumptions about the data that aid in performance. Each dimension is an equivalent entry point into the fact table, and this symmetrical structure allows effective handling of complex queries. Query optimization for star join databases is simple, predictable, and controllable.

3. **Extensibility:** Dimensional models are extensible and easily accommodate unexpected new data. Existing tables can be changed in place either by simply adding new data rows into the table or executing SQL alters table commands. No queries or other applications that sit on top of the Warehouse need to be reprogrammed to accommodate changes. Old queries and applications continue to run without yielding different results. But in normalized models each modification should be considered carefully, because of the complex dependencies between database tables.

3.1.6.1 Dimensional Modelling Process

Dimensional modelling gets its name from business dimensions we need to incorporate into logical data model. It is a logical design technique to structure the business dimensions and metrics that are analysed along these dimensions (Ralph Kimball). Dimensional modelling is a new name for an old technique for making database simple and understandable. When a database can be visualized as a "cube" for three, four or five dimensions, people can imagine slicing and dicing that cube along each of these dimensions (Ralph Kimball). The model has also provided high performance for queries and analysis. Dimensional modelling is deferent way to view and interrogate data in a database. This view may be used in DSS in conjunction with data mining tasks.

Dimensional modelling gives us the ability to visualize the data. A dimension is a collection of logically related attributes and is viewed as an axis for modelling the data. The dimension could be divided into many different granularities: decade, year, month, day, and hour. A dimensional model contains two entity types: facts and dimensions. We keep the measurements of a given business process in entities called fact tables in the relational world, or measure groups in analysis services. In its classic form, a relational fact table contains foreign keys and numeric fact fields that measure the business process. Essentially, a fact table is a normalized structure for efficient storage and fast access of large, relational data sets. Before creating a dimensional model, we should understand the basic objects that we use to create dimensional models:

1. **Fact tables and entities**
2. **Dimension tables and entities**
3. **Hierarchies**
4. **Outriggers, and**
5. **Measures**

- **Fact tables and entities:** A *fact table* or a *fact entity* is a table or entity in a star or snowflake schema that stores measures that measure the business, such as sales, cost of goods, or profit.
- **Dimension tables and entities:** A *dimension table* or *dimension entity* is a table or entity in a star, snowflake, or starflake schema that stores details about the facts. For example, a Time dimension table stores the various aspects of time such as year, quarter, month, and day.
- **Hierarchies:** A *hierarchy* is a many-to-one relationship between members of a table or between tables. A hierarchy basically consists of different levels, each corresponding to a dimension attribute.
- **Outriggers:** An *outrigger* is a dimension table or entity that is joined to other dimension tables in a star

schema. Outriggers are used when a dimension table is snowflaked.

- **Measures**: *Measures* define a measurement attribute and are used in fact tables. You can calculate measures by mapping them directly to a numerical value in a column or attribute. An aggregation function summarizes the value of the measure for dimensional analysis.

Star Schema: A star schema model can be depicted as a simple star: a central table contains fact data and multiple tables radiate out from it, connected by the primary and foreign keys of the database. In a star schema implementation, Warehouse Builder stores the dimension data in a single table or view for all the dimension levels.

A star schema is the simplest form of a dimensional model, in which data is organized in to *facts and dimensions*. A fact is an event that is counted or measured, such as a production of performance. A dimension contains reference information about the fact. A star schema is diagrammed by surrounding each fact with its associated dimensions. The resulting diagram resembles a star. Star schemas are optimized for querying large data sets and are used in data warehouses and data marts to support OLAP cubes, business intelligence and analytic applications, and ad-hoc queries. Within the data warehouse or data mart, a dimension table is associated with a fact table by using a foreign key relationship. The dimension table has a single primary key that uniquely identifies each member record (row). The fact table contains the primary key of each associated dimension table as a foreign key. Combined, these foreign keys form a multi-part composite primary key that uniquely identifies each member record in the fact table. The fact table also contains one or more numeric measures.

The star schema supports rapid aggregations such as count, sum, and average of many fact records, and these aggregations can be easily filtered and grouped by the dimensions. A star schema may be partially normalized or snowflaked, with related information stored in multiple related dimension tables, to support specific data warehousing needs.

Snowflake Schema: The snowflake schema represents a dimensional model which is also composed of a central fact table and a set of constituent dimension tables which are further normalized into sub-dimension tables. In a snowflake schema implementation, warehouse builder uses more than one table or view to store the dimension data. Separate database tables or views store data pertaining to each level in the dimension. In snowflake schema, dimensions are normalized into multiple related tables, where in star schema dimensions are normalized with each dimension represented by a single table.

A snowflake schema describes a popular data model for relational data warehouses. Similar to a star schema, a snowflake schema contains fact and dimension tables, with the fact table storing facts about business transactions and some columns of the fact table containing foreign key references to the dimension tables. Snowflake schemata differ from star schemata in their level of normalization; the dimension tables of a snowflake schema are typically normalized to third normal form (3NF) or higher.

As a result, the data for a given dimension is spread out over multiple tables, and a diagram of the database resembles a snowflake. While a snowflake schema can result from normalization of the dimension tables, it can also arise through dimension table reuse, particularly when there are relationships between dimension tables.

3.1.6.2 Dimensional Modelling Steps

The dimensional model is built on a star-like schema, with dimensions surrounding the fact table. To build the schema, the following design model is used: (Ralph Kimball).

1. **Choose the Business Process**
2. **Declare the Grain (Attributes)**
3. **Identify the Dimensions**
4. **Identify the Fact**

Choose the Business Process: The process of dimensional modelling builds on a 4-step design method that helps to ensure the usability of the dimensional model and the use of the data warehouse. The basics in the design build on the actual business process which the data warehouse should cover. Therefore, the first step in the model is to describe the business process which the model builds on. To describe the business process, one can choose to do this in plain text or use basic Business Process Modelling Notation (BPMN) or other design guides like the *Unified Modelling Language* (UML).

Declare the Grain (Attributes): After describing the Business Process, the next step in the design is to declare the grain of the model. The grain of the model is the exact description of what the dimensional model should be focusing on. This could for instance be "An individual line item on a customer slip from a retail store". To clarify what the grain means, you should pick the central process and describe it with one sentence. Furthermore, the grain (sentence) is what you are going to build your dimensions and fact table from. One might find it necessary to go back to this step to alter the grain due to new information gained on what the model is supposed to be able to deliver.

Identify the Dimensions: The third step in the design process is to define the dimensions of the model. The dimensions must be defined within the grain from the second step of the 4-step process. Dimensions are the foundation of the fact table, and are where the data for the fact table is collected. Typically, dimensions are nouns like date, store, inventory etc. These dimensions are where all the data is stored. For example, the date dimension could contain data such as year, month and weekdays.

Identify the Facts: After defining the dimensions, the next step in the process is to make keys for the fact table. This step is to identify the numeric facts that will populate each fact table row. This step is closely related to the business users of the system, since this is where they get access to data stored in the data warehouse. Therefore, most of the fact table rows are numerical, additive figures such as quantity or cost per unit.

3.1.6.3 Dimensional Normalization

Dimensional normalization or snowflaking removes redundant attributes, which are known in the normal, flatten de-normalized dimensions. Dimensions are strictly joined together in sub-dimensions. Snowflaking has an influence on the data structure that differs from many philosophies of data warehouses (Matteo Golfarelli, Stefano Rizzi). Single data (fact) table surrounded by multiple descriptive tables. Developers often don't normalize dimensions due to several facts (Ralph Kimball, Margy Ross) Normalization makes the data structure more complex that resultant:

1. Performance can be slower, due to the many joins between tables
2. The space savings are minimal

3. The use of bitmap indexes can't be done
4. Query Performance, 3NF databases suffer from performance problems when aggregating or retrieving many dimensional values that may require analysis.

If you are only going to do operational reports then you may be able to get by with 3NF because your operational user will be looking for very fine grain data. There are some arguments on why normalization can be useful.

3.1.7 The Data Warehouse Design Process

Building a dimensional data warehouse is a process of matching the needs of the user community to the realities of the available data. There is a set of nine major (Ralph Kimball) decision points in the database design that will be driven by the needs of user community and the realities of the data. Ralph Kimball gives the methodology by briefly enumerating the nine decision points of a complete database design. The nine decision points of a complete database design for a dimensional data warehouse consists of deciding on the following:

1. The process, and hence the identity of the fact tables
2. The grain of each of the fact table
3. The dimensions of each fact table
4. The, facts including pre-calculated facts
5. The dimensional attributes with complete description and proper terminology
6. How to track slowly changing dimensions
7. The aggregation, heterogeneous dimensions, mini-dimensions, queries modes and other decisions
8. The historical duration of the database

9. The urgency with which the data is extracted and loaded into the data warehouse

Ralph Kimball recommended that these nine decisions be made in given order. The following are the typical steps involved in the data warehousing project cycle:

1. **Requirement Gathering**
2. **Physical Environment Setup**
3. **Data Modelling**
4. **ETL (Extraction, Transformation, Loading)**
5. **OLAP Cube Design**
6. **Front End Development**
7. **Report Development**
8. **Performance Tuning**
9. **Query Optimization**
10. **Quality Assurance**
11. **Rolling out to Production**
12. **Production Maintenance**
13. **Incremental Enhancements**

Each of listed above represents a typical data warehouse design phase & has several sections:

- **Task Description:** This section describes what typically needs to be accomplished during this particular data warehouse design phase.
- **Deliverables:** Typically, at the end of each data warehouse task, one or more documents are produced that fully describe the steps and results of that particular task. This

is very important for consultants to communicate their results to clients.

- **Possible Pitfalls:** Things to watch out for. Some of them obvious, some of them not so obvious.

The steps & sections recommended above by *Ralph Kimball* is slightly modified and customized as per the requirement of this particular work, author is obliged for his contribution to this research work. The customized data warehouse design process is explained in following sections.

3.1.7.1 Requirement Gathering (Step -1)

Task Description: The first thing that the project team should engage in is gathering requirements from end users. Because end users are typically not familiar with the data warehousing process or concept, the help of the business sponsor is essential. Collections of agricultural raw facts are obviously first step in proposed work. Indian agriculture is often more typical due to its varieties of crops, land and environmental issues. It is a challenging job that required proper survey and observation. This observation can be categorized in three basic issues: land, crops and environment.

The primary goal of this phase is to identify what constitutes as a success for this particular phase of the data warehouse project. In particular, end user reporting analysis requirements are identified, and the project team will spend the remaining period of time trying to satisfy these requirements. For any agricultural data warehouse '**soil-type**' is the first core issue that are classified on the basis of geographical distribution in India. The '**climate**' is taken as second important attributes for this work and '**fertilizers & minerals**' as third important input. As it is common knowledge that due to lack of 'fertilizers &

minerals ′, a big part of agricultural land is not found suitable for cropping. This land can be re-utilized by use of fertilizers and reformatting. These fertilizers some time generated by crops itself, and so the sequence of crops for a particular land is very significant to include in survey. of fertilizers and minerals in soil is clay and climate conditions.

Deliverables:

- A list of reports/cubes to be delivered to the end users by the end of this current phase.
- An updated project plan that clearly identifies resource loads and milestone delivery dates.

Possible Pitfalls:

This phase often turns out to be the trickiest phase of the data warehousing implementation. The reason is that because data warehousing includes data from multiple sources spanning many different areas and government departments. There are often some tremendous problems in account of willingness of information sharing by the government department. As a result of unwillingness of certain department to release data or to participate in the data warehousing requirement gathering, the effort either never gets off the ground, or could not start in the direction originally defined.

3.1.7.2 Physical Environment Setup (Step -2)

Task Description: Once the requirements are somewhat clear, it is necessary to set up the physical servers and databases. At a minimum, it is necessary to set up a development environment and a production environment. There are also many data warehousing projects where there are three environments: Development, Testing, and Production.

Data warehouse is processed on informational data, rather than operational data, and so these raw facts required processing for converting operational data into informational one. The data is first classified in different environment attributes and then placed in multidimensional schema. This includes many databases prospective to examining information retrieval and data mining concepts, efficiency and scalability. It is not enough to simply have different physical environments set up. The different processes such as ETL, OLAP Cube, are need to be set up properly for each environment. It is best for the different environments to use distinct application and database servers. In other words, the development environment will have its own application server and database servers, and the production environment will have its own set of application and database servers.

Having different environments is very important because it allow the data warehousing team to examine the data without impacting the production environment.

Deliverables:

- Hardware / Software setup document for all of the environments, including hardware specifications, and scripts / settings for the software.

Possible Pitfalls:

In order to save, often data warehousing teams will decide to use only a single database and a single server for the different environments. This is problematic for the following reasons:

- Sometimes it is possible that the server needs to be rebooted for the development environment. Having a separate development environment will prevent the production environment from being impacted by this.

- There may be interference when having different database environments on a single box. For example, having multiple long queries running on the development database could affect the performance on the production database.

3.1.7.3 Data Modelling (Step-3)

Task Description: This is a very important step in the data warehousing project. It is the foundation of the data warehousing system. A good data model will allow the data warehousing system to grow easily, as well as for providing good performance. In data warehousing project, the logical data model is built based on user requirements, and then it is translated into the physical data model. Part of the data modelling exercise is often the identification of data sources.

Deliverables:

- Identification of data sources.
- Logical data model.
- Physical data model.

Possible Pitfalls:

It is essential to have a subject-matter expert as part of the data modelling team. This person can be an outside consultant or can be someone in-house who has extensive experience in the industry. Without this person, it becomes difficult to get a definitive answer on many of the questions, and the entire project gets dragged out. In our study three major attributes soil type, crops and climate (calendar month) is chosen as dimensional of multidimensional schema.

3.1.7.4 ETL (Extraction, Transformation, Loading) (Step- 4)

Task Description: The ETL (Extraction, Transformation, Loading) process is probably longest step that need extra time to develop, and this take half of efforts and time of data warehouse implementation cycle. The reason for this is that it takes time to get the source data, understand the necessary attributes, understand the agricultural process, and identify the most suitable logical and physical data models.

Deliverables:

- Data Mapping Document
- Need of extraction and Data Mining tool

Possible Pitfalls:

There is a tendency to give this particular phase little development time than it requires. This can spoil the objective of the project because end users (farmers) will usually tolerate less formatting, longer time to run reports, less functionality, and fewer delivered reports. A second common problem is that some people make the ETL process more complicated than necessary. When this happens, ETL performance suffers, and often so does the performance of the entire data warehousing system.

3.1.7.5 OLAP Cube Design (Step- 5)

Task Description: Usually, the design of the OLAP cube can be derived from the requirement Gathering phase. More often, data warehouse designer has some idea on what they want, but it is difficult for them to specify the exact analysis they want to see. It is usually a good idea to include enough information so that they feel like they have gained something through the data warehouse. Remember that data warehousing is an iterative process no one can ever meet all the requirements all at once. In this study the three major attributes are placed in the cube, and a correspondent Star schema is built that shows the relation among the selected attributes. Further this star schema is extended into snowflake schema that clearly represents the relationship among other factors effecting attributes.

Deliverables:

- The OLAP cube and its dimensions as the major attributes or decision factor.
- OLAP cube-based star schema and snowflake schema.

Possible Pitfalls:

Make sure your OLAP cube-building process is optimized. It is common error to in appropriate selection of dimensions of the OLAP cube. A balanced number of records in each dimension is recommended. However, it is not always possible neither necessary rule. An equal distribution of the data population reduced data fetching time and for increase refresh rate of OLAP cube. So, it is worthwhile to deal with OLAP cube generation paths in order to ensure optimal performance.

3.1.7.6 Front End Development (Step- 6)

Task Description: The front-end development is an important part of a data warehousing initiative, because; visualization of the report is *'the final fruit'* of all the efforts made by data warehouse designers. If the end users are not able to visualize the reports, the data warehousing efforts brings zero value to them. The selection of a front-end deployment methodology is crucial issue for many reasons.

The most important thing is that the reports should need to be delivered over the web, so it can be accessed from anywhere. So, whatever strategy one pursues, make sure the ability to deliver over the web is a must.

The front-end options range from an internal front-end development using scripting languages such as ASP, PHP, or

Perl, to off-the-shelf products such as Seagate Crystal Reports, to the more high-level products such as Actuate. In addition, many OLAP vendors offer a front-end on their own. When choosing vendor tools, make sure it can be easily customized to suit the enterprise, especially the possible changes to the reporting requirements of the enterprise. Possible changes include not just the difference in report layout and report content, but also include possible changes in the back-end structure.

In our study the GIS is one of the suitable tools to visualized data to the farmers. In this study the use of ArcInfo or similar software is recommended because it can store different information each on the different layers of the map. The GIS tools are easy to navigate and understand by the farmers, and also it can be easily connected to the web application.

Deliverables:

- Front End Deployment Documentation

Possible Pitfalls:

Just remember that the end users do not care how complex or how technologically advanced your front-end infrastructure is. All they care is that they receive their information in a timely manner and in the way they specified

3.1.7.7 Report Development (Step- 7)

Task Description: Report specification typically comes directly from the requirements phase. To the end user, the only important point with the data warehousing system is the reports they see. So, report development plays a very important role in determining the success of the data warehousing project. There are several points the data warehousing team need to pay attention to before releasing the report.

1. **User customization:** Do users need to be able to select their own metrics? And how do users need to be able to filter the information? The report development process needs to take those factors into consideration so that users can get the information they need in the shortest amount of time possible.

2. **Report delivery:** What report delivery methods are needed? In addition to delivering the report to the web front end, other possibilities include delivery via email, via text messaging, or in some form of spreadsheet. There are reporting solutions in the marketplace that support report delivery as a flash file. Such flash file essentially acts as a mini-cube, and would allow end users to slice and dice the data on the report without having to pull data from an external source.

3. **Access privileges:** Special attention needs to be paid to who has what access to what information. A sales report can show 8 metrics covering the entire company to the company CEO, while the same report may only show 5 of the metrics covering only a single district to a District Sales Director.

Report development does not happen only during the implementation phase. After the system goes into implementation, there will certainly be requests for additional reports. These types of requests generally fall into two broad categories:

a. Data is already available in the data warehouse. In this case, it should be fairly straightforward to develop the new report into the front end. There is no need to wait for a major production push before making new reports available.

b. Data is not yet available in the data warehouse. This means that the request needs to be prioritized and put into a future data warehouse.

Deliverables:

- Report Specification Documentation.
- Reports set up in the front end / reports delivered to user's preferred channel.

Possible Pitfalls:

Make sure the exact definitions of the report are communicated to the users. Otherwise, user interpretation of the report can be erroneous.

3.1.7.8 Performance Tuning (Step- 8)

Task Description: There are three major areas where a data warehousing system can use a little performance tuning:

ETL - Given that the data load is usually a very time-consuming process and that data warehousing-related batch jobs are typically of lower priority. A data warehousing system that has its ETL process finishing right on-time is going to have a lot of problems simply because often the jobs do not get started on-time due to factors that is beyond the control of the data warehousing team. As a result, it is always an excellent idea for the data warehousing group to tune the ETL process as much as possible.

Query Processing - Sometimes, especially in a ROLAP environment or in a system where the reports are run directly against the relationship database, query performance can be an issue. *Ralph Kimball* has presented a number of query optimization ideas.

Report Delivery - It is also possible that end users are experiencing significant delays in receiving their reports due to factors other than the query performance. For example, network traffic, server setup, and even the way that the front-end was built. It is important for the data warehouse team to look into these areas for performance tuning.

Deliverables

- Performance tuning document - Goal and Result

Possible Pitfalls:

Make sure the development environment is on the same track of the production environment as close as possible.

3.1.7.9 Query Optimization (Step- 9)

The SQL query optimization is a major issue because it not only effect performance of the data warehouse but also leads to issues related to data corruption and the accessing speed of repository. sooner or later. Having long-running queries not only consumes system resources it also makes the server and application run slowly and that may lead to table locking. In order to optimize the ad-hoc queries the Data Mining Query Language (DMQL) is proposed for this warehousing project. As we may aware that the DMQL is designed to work closely with the programming languages. Proposed DMQL is in fact an Object Query Language (OQL), which allows to access background information of objects, which are actually attributes in multidimensional schema. The following are some important consideration to optimize user's query.

1. **Understand how your database is executing your query**

The first step to query optimization understands what the database is doing. Different databases have different commands for this. For example, in MySQL, we can use "EXPLAIN [SQL Query]" keyword to see the query plan. In Oracle, we can use "EXPLAIN PLAN FOR [SQL Query]" to see the query plan.

2. Retrieve as little data as possible

The more data returned from the query, the more resources the database needs to expand to process and store these data. For example, if you only need to retrieve one column from a table, do not use 'SELECT *'.

3. Store intermediate results

Sometimes logic for a query can be quite complex. Often, it is possible to achieve the desired result through the use of sub-queries, inline views, and UNION-type statements. For those cases, the intermediate results are not stored in the database, but are immediately used within the query. This can lead to performance issues, especially when the intermediate results have a large number of rows.

The way to increase query performance in those cases is to store the intermediate results in a temporary table, and break up the initial SQL statement into several SQL statements. In many cases, you can even build an index on the temporary table to speed up the query performance even more. Granted, this adds a little complexity in query management (i.e., the need to manage temporary tables), but the speedup in query performance is often worth the trouble.

Ralph Kimball has given several specific query optimization strategies.

- **Use Index:** Using an index is the first strategy one should use to speed up a query. In fact, this strategy is so important that index optimization is also discussed.

- **Aggregate Table:** Pre-populating tables at higher levels so less amount of data needs to be parsed.
- **Vertical Partitioning:** Partition the table by columns. This strategy decreases the amount of data a SQL query needs to process.
- **Horizontal Partitioning:** Partition the table by data value, most often time. This strategy decreases the amount of data a SQL query needs to process.
- **De-normalization:** The process of de-normalization combines multiple tables into a single table. This speeds up query performance because fewer table joins are needed
- **Server Tuning:** Each server has its own parameters, and often tuning server parameters so that it can fully take advantage of the hardware resources can significantly speed up query performance.

3.1.7.10 Quality Assurance (Step- 10)

Task Description: Once the development team declares that everything is ready for further testing, the QA team takes over. Usually, the Quality Assurance team members will know little about data warehousing, and some of them may even resent the need to have to learn another tool or tools. Sometimes the Quality Assurance process is overlooked. Some time it may took extra months to complete the project. To avoid any such, delay a preparing a timeline based work-plan is considered as the best practice.

Deliverables:

- Quality Assurance Test Plan
- Quality Assurance Verification

Possible Pitfalls:

As mentioned above, usually the Quality Assurance team members know little about data warehousing, and some of them may even resent the need to have to learn another tool or tools. Make sure the QA team members get enough education so that they can complete the testing themselves.

3.1.7.11 Rollout to Production (Step- 11)

Task Description: After successful implementation of the QA phase, it is time for the data warehouse system to go live. However, making all data warehouse available online at once is not easy task. Depending on the number of end users, it sometimes takes a week or more to bring everyone online. Fortunately, nowadays most end users access the data warehouse over the web, making going production sometimes as easy as sending out an URL via email.

Deliverables:

- Delivery of the data warehousing system to the end users.

Possible Pitfalls:

This phase addresses the education and training concern of the end user. There is nothing more frustrating to spend several months to develop and QA the data warehousing system, only to have little usage because the users are not properly trained. Regardless of how intuitive or easy the interface may be; it is always a good idea to send the users to at least a one-day course to let them understand what they can achieve by properly using the data warehouse.

3.1.7.12 Production Maintenance (Step- 12)

Task Description: Once the data warehouse goes production, it needs to be maintained. Tasks as such regular backup and crisis management becomes important and should be planned out. This serves two purposes:

1. To capture any runaway requests so that they can be fixed before slowing the entire system down, and
2. To understand how much users are utilizing the data warehouse for return-on-investment calculations and future enhancement considerations.

Deliverables:

- Consistent availability of the data warehousing system to the end users.

Possible Pitfalls:

Usually by this time most of the developers will have left the project, so it is essential that proper documentation is left for those who are handling production maintenance.

3.1.7.13 Incremental Enhancements (Step- 13)

Task Description: Once the data warehousing system goes live, there are often needs for incremental enhancements. I am not talking about a new data warehousing phases, but simply small changes that follow the business itself.

Deliverables:

- Change management documentation
- Actual change to the data warehousing system

3.2 Data Mining

The first International Conference on Knowledge Discovery and Data Mining (KDD) was held in 1995, and there are a variety of definitions of data mining. A concise definition that captures the essence of data mining is:

Extracting useful information from large datasets (Hand et al., 2001)

Data mining is the process of exploration and analysis, by automatic or semi-automatic means, of large quantities of data in order to discover meaningful patterns and rules. (Berry and Linoff, 1997, 2000).

Another definition comes from Gartner Group, the information technology research firm (from their Web site, January 2004):

Data mining is the process of discovering meaningful new correlations, pattern and trends by sifting through large amount of data stored in repositories, using pattern recognition technologies as well as statistical and mathematical techniques.

By combining above two definitions, we can conclude:

Data mining, also called knowledge discovery in databases is the process of discovering interesting and useful patterns and relationships in large volumes of data.

The field combines tools from statistical and artificial intelligence, with database management to analyse large digital collection of records, known as data-sets. The overall goal of the data mining process is to extract information from a data set and transform it into an understandable structure for further use. Data mining automates the detection of relevant patterns in a database, using defined approaches and algorithms to look into current and historical data that can then be analysed to predict future trends. Because data mining tools predict future trends and behaviours by reading through databases for hidden patterns, they allow organizations to make proactive, knowledge-driven decisions and answer questions that were previously too time-consuming to resolve.

Data Mining is not New: Manual statisticians and similar approaches have been used to review data for providing business projections since many years. However, development in data mining techniques, have equipped organizations to collect, analyse, and access data in entirely new way. The major changes occurred in the way of data acquisition, and data collection. Earlier, companies made the transition from ledgers and other paper-based records that further stored into computer systems. The decision maker and managers had to wait for their staff to combine the data sets and analyse it for business queries such as, "how well the business was performing?" or "how current performance periods compared with previous periods". As companies started collecting and saving basic data in computers, they were able to start answering detailed questions quicker and with more ease.

Changes in Data Access Methods: Over the last 30 years, there has been greater advancement and integration of techniques, particularly in acquiring of heterogeneous data sets and its management. The introduction of microcomputers and networks, and the evolution of network protocols, and use of IoT *(Internet of Things)* methodologies, have enable data to be moved seamlessly among programs and other machines. It, allows companies to link common database queries together. The development of data warehousing and decision support systems, has enabled companies to extend queries from "What was happen? " to "What is likely to happen to, and why?"

Vast Amount of Information: Now days, organizations have more information their disposal. With extensively growth in amount of data collection and continuously use of new ways of data collection, the data mining is not uncommon approach to investigate trends and after quick processing of large volumes of data. Users can determine the outcome of the data analysis by the parameters they chose, thus providing additional value to business strategies and initiatives. It is important to note that without these parameters, the data mining program will generate all permutations or combinations irrespective of their relevance. In this study we have chosen *association rule* and *decision tree* for future prediction about farmer's query against prosed agricultural data warehouse.

Internal Auditing: As we know that the data mining programs has the lack of having capabilities like *human intuition* that can recognize the difference between a relevant and an irrelevant data correlation, users need to review the results of mining exercises to ensure results provide needed information. For example, knowing that people who default on loans usually give a false address might be relevant, whereas knowing they have blue eyes might be irrelevant. Auditors, therefore, should monitor whether sensible and rational decisions are made on the basis of data mining exercises, especially where the results of such exercises are used as input for other processes or systems. Auditors also need to consider the different security aspects of data mining programs and processes. A data mining exercise might reveal important customer information that could be exploited by an outsider who hacks into the rival organization's computer system and uses a data mining tool on captured information.

3.2.1 Data Mining Tasks

The actual data mining task is the automatic or semi-automatic analysis of large quantities of data to extract previously unknown and interesting patterns such as groups of data records also known as *'cluster analysis'*, unusual records or *'anomaly detection'* and dependencies or *'association rule mining'*. This usually involves using database techniques such as spatial indices. These patterns can then be seen as a kind of summary of the input data, and may be used in further analysis or, for example, in machine learning and predictive analytics. For example, the data mining step might identify multiple groups in the data, which can then be used to obtain more accurate prediction results by a decision support system. Data mining involves six common classes of tasks (Fayyad, Usama 2008).

- **Anomaly detection (Outlier/change/deviation detection):** The identification of unusual data records, that might be interesting or data errors that require further investigation.
- **Association rule learning (Dependency modelling):** Searches for relationships between variables. For example, a supermarket might gather data on customer purchasing habits. Using association rule learning, the supermarket can determine which products are frequently bought together and use this information for marketing purposes. This is sometimes referred to as market basket analysis.
- **Clustering:** is the task of discovering groups and structures in the data that are in some way or another "similar", without using known structures in the data.
- **Classification:** is the task of generalizing known structure to apply to new data. For example, an e-mail program might attempt to classify an e-mail as 'legitimate' or as 'spam'.
- **Regression:** Attempts to find a function which models the data with the least error.
- **Summarization:** providing a more compact representation of the data set, including visualization and report generation.

3.2.1.1 Supervised and Unsupervised Learning

The data mining techniques is theoretically categorized in two methods, *supervised learning* and *unsupervised learning*.

Supervised learning algorithms are those used in classification and prediction. One must have information like; data available in which format the value of the outcome of interest is known. These training data are the data from which the classification or prediction algorithm learns or is trained, about the relationship between predictor variables and the outcome variable. Once the algorithm has learned from the training data, it is then applied to another sample of data (the validation data) where the outcome is known, to see how well it does in comparison to other models. If many different models are being tried out, it is prudent to save a third sample of known outcomes (the test data) to use with the model finally selected to predict how well it will do. The model can then be used to classify or predict the outcomes is unknown. Simple liner regression analysis is an example of supervised learning. The Y variable is known outcome variable and the X variable is a predictor variable. A regression line is drawn to minimize the sum of squared deviations between the actual Y values and the values predicted by this line. The regression line can then be used to predict Y values for new values of X for which we do not know the Y value.

Unsupervised learning algorithms are those used where there is no outcome variable to predict of classify. Hence, there is no 'learning' from case where such an outcome variable is known. Association rules, data reduction methods, and clustering techniques are all unsupervised learning methods.

3.2.2 Steps of Data Mining

There are various steps that are involved in mining data

1. **Data Integration:** First of all, the data are collected and integrated from all the different sources.

2. **Data Selection:** We may not all the data we have collected in the first step. So, in this step we select only those data which we think useful for data mining.

3. **Data Cleaning:** The data we have collected are not clean and may contain errors, missing values, noisy or inconsistent data. So, we need to apply different techniques to get rid of such anomalies.

4. **Data Transformation:** The data even after cleaning are not ready for mining as we need to transform them into forms appropriate for mining. The techniques used to accomplish this are smoothing, aggregation, normalization etc.

5. **Data Mining:** Now we are ready to apply data mining techniques on the data to discover the interesting patterns. Techniques like clustering and association analysis are among the many different techniques used for data mining.

6. **Pattern Evaluation & Knowledge Presentation:** This step involves visualization, transformation, removing redundant patterns etc. from the patterns we generated.

7. **Decisions / Use of Discovered Knowledge:** This step helps user to make use of the knowledge acquired to take better decisions.

The business value of unstructured data can be left unexploited when data mining is executed with an inadequate, static data model designed for structured data. The static business models and static data models are used to model dynamic business processes, particularly at the point of interaction. For example, virtually every customer relationship management system has a manual classification scheme that is meant to be used by the service agent to classify the nature of the customer interaction.

This approach has two major flaws. First, as soon as the classification scheme is published, it is out of date, because interactions with the customers are unpredictable and continually changing. Second, even if the classification scheme was representative of the customer interactions, it is unreasonable to expect any service agents to classify their interactions with their customers in a consistent way and with high quality. This very often makes such classification data completely useless, or, more dangerously, misleading. This issue is true throughout the business ecosystem where unstructured information exists.

An adaptable data model is critical when incorporating unstructured information mining. One problem commonly encountered is that the analysis typically leads to more questions. In business intelligence or data mining, if the data model is not designed to handle the new question, the data model must be modified and the data manipulated and reloaded, which is often a difficult and cumbersome process, many times taking months. This problem is compounded even more with unstructured information because of its very nature. It is important to be able to add or enhance existing taxonomies, classifications, or extracted data as the information leads you through the discovery process. Additionally, it is important to combine the right mix of algorithmic assistance with domain expertise. We have found that most people naively want to push a button and magically receive the answer. However, we have come to the conclusion that, particularly with unstructured information, the analytical process cannot be fully automated. Even for structured information, it typically needs an analyst in the loop to interpret the results. For unstructured information, an analyst is required in the loop to help guide the process with a combination of algorithmic assistance, a useful set of metrics to assist in the interpretation, and the analyst's domain expertise. The key is to make efficient use of this expensive and scarce resource, not to eliminate it entirely.

-Scott Spangler and Jeffrey Kreulen, Introduction to Mining the Talk: Unlocking the Business Value in Unstructured Information, Jun 24, 2008

3.2.3 Data Mining Tools

Various data mining tools are under practice in the marketplace, each of them has its own strengths and weaknesses. Internal data auditors need to keep an eye on different types of data mining tools available and recommend the purchase of a tool that fulfil the current needs. This should be estimate as early as possible in the project development lifecycle, perhaps during the feasibility study. Most of the data mining tools can be classified into one of three categories: traditional data mining tools, dashboards, and text-mining tools. The detail description of each is given below.

Traditional Data Mining Tools: Traditional data mining is based on establishing data patterns and trends by using an expert and complex algorithms and data mining tools. Some of these tools are installed on the desktop to continuously monitor the change in data patterns and highlight the trends and others important information that residing outside the database. The majority of such tools are available for Windows and UNIX operating systems; however, some are specialized for one operating system only. In addition, most of the data mining tools are able to handle any data using online analytical processing or a similar technology.

Dashboards: Dashboards are installed in computers to monitor information flow in a database, dashboards capture the data changes that and reflect and updates onscreen, often in the form of a chart or table that enable to the user to visualize live performance of the business. For the purpose of this study the historical agriculture data can be referenced, to observe the change in trends (e.g., increase in production of crop from the same period last year). This functionality makes dashboards easy to use and particularly appealing to farmers who wish to have an overview of the crop performance.

3.2.3.1 Data Mining Strategies

When evaluating data mining strategies, the database manager may decide to purchase several tools for different purposes, rather than acquire one common tool that meets all needs. Organizations that wish to use data mining tools can purchase mining programs designed for existing software and hardware platforms, which can be integrated into new products and systems as they are brought online, or they can build their own custom mining solution. For instance, feeding the output of a data mining exercise into another computer system, such as a neural network, is quite common and can give the mined data more value. This is because the data mining tool gathers the data.

3.2.4 Text-Mining Tools

The third type of data mining tool is *text-mining.* It has ability to mine data from different kinds of text, like from Microsoft Word and Acrobat PDF documents to simple text files. These tools scan the content and convert the selected data into a standard format that is compatible with the database. It provides an easy and convenient way of accessing data without the need to open different application files. Scanned content can be unstructured or structured.

3.2.5 Data Mining: Functions and Methods

Data mining is an iterative process, that defines progress through observe changes in data, through either automatic or manual methods. Data mining is most useful for exploratory analysis scenario in which there are no predetermined notions about what will be the *interesting outcome*. Data mining is use to explore new, valuable, and nontrivial information in large volumes of data. It is a cooperative effort of humans and computers, and the best results could be achieved by balancing the knowledge of human experts in describing problems and goals with the search capabilities of computers.

In practice, the two primary goals of data mining tend to be prediction and description, that originate term *predictive data mining* and *descriptive data mining* respectively. Prediction involves using some variables or fields in the data set to predict unknown or future values of other variables of subject interest. On the other hand, the description, focuses on finding patterns describing the data that can be interpreted by humans. More clearly the both can be understood as:

1. Predictive data mining, which produces the model of the system described by the given data set, and
2. Descriptive data mining, which produces new, nontrivial information based on the available data set.

The predictive data mining is aiming is to produce a conceptual or mathematical model, which can be used to perform classification, prediction, estimation, or other similar tasks. On the other, descriptive data mining has the objective is to gain an understanding of the analysed system by unveiling the patterns and correlations in attributes of large data sets. For mining of agricultural data warehouse, we can use a balanced mix of predictive and descriptive data mining.

Data mining for agricultural data warehouse has its origins in collecting data from heterogeneous sources and fields. The process of mining such complex and vast data warehouse is based on two most important discipline statistics and machine learning. Statistics has its roots in mathematics, and therefore, there has been an emphasis on mathematical results that caries few complexities especially, when interpreting the results. It is difficult to say that the result of such statistical-analysis will be sensible on theoretical grounds before testing it in practice. In contrast, the machine-learning community has its origins very much in computer practice. This has led to a practical orientation, a willingness to test something out to see how well it performs, without waiting for a formal proof of effectiveness. The various functions of data mining are as follows:

1. **Classification** - discovery of a predictive learning function that classifies a data item into one of several predefined classes.
2. **Regression** - discovery of a predictive learning function, which maps a data item to a real-value prediction variable.
3. **Clustering** - a common descriptive task in which one seeks to identify a finite set of categories or clusters to describe the data.
4. **Summarization** - an additional descriptive task that involves methods for finding a compact description for a set (or subset) of data.
5. **Dependency Modelling** - finding a local model that describes significant dependencies between variables or between the values of a feature in a data set or in a part of a data set.

6. **Change and Deviation Detection** - discovering the most significant changes in the data set.

3.2.6 Data Mining Techniques

A good way to apply advanced data mining techniques is to have a flexible and interactive data mining tool that is fully integrated with a database or data warehouse. Using a tool that operates outside of the database or data warehouse is not as efficient. Using such a tool will involve extra steps to extract, import, and analyse the data. When a data mining tool is integrated with the data warehouse, it simplifies the application and implementation of mining results. Furthermore, as the warehouse grows with new decisions and results, the organization can mine best practices continually and apply them to future decisions. Regardless of the technique used, the real value behind data mining is modelling the process of building a model based on user-specified criteria from already captured data. Once a model is built, it can be used in similar situations where an answer is not known.

Each of these approaches brings different advantages and disadvantages that need to be considered prior to their use. Neural networks, which are difficult to implement, require all input and resultant output to be expressed numerically, thus needing some sort of interpretation depending on the nature of the data-mining exercise. The decision tree technique is the most commonly used methodology, because it is simple and straightforward to implement. Finally, the nearest-neighbour method relies more on linking similar items and, therefore, works better for extrapolation rather than predictive enquiries. We have chosen some of most commonly used techniques that include artificial neural networks, decision trees, and the nearest-neighbour method. Each of these technique analyses data in different ways:

3.2.6.1 Decision Tree (DT)

Decision Trees are tree-shaped structures that represent decision sets. These decisions generate rules, which then are used to classify data. Decision trees are the favoured technique for building understandable models. Auditors can use them to assess, for example, whether the organization is using an appropriate cost-effective marketing strategy that is based on the assigned value of the customer, such as profit. It is a decision support tool that uses a tree-like graph or model of decisions and their possible consequences, including chance event outcomes, resource costs, and utility. It is one way to display an algorithm. Decision trees are commonly used in operations research, specifically in decision analysis, to help identify a strategy most likely to reach a goal.

In decision analysis a decision tree and the closely related influence diagram is used as a visual and analytical decision support tool, where the expected values (or expected utility) of competing alternatives are calculated. **A decision tree consists of 3 types of nodes.**

1. **Decision nodes**
2. **Chance nodes**
3. **End nodes**

Decision trees are commonly used in operations research, specifically in decision analysis, to help identify a strategy most likely to reach a goal. If in practice decisions have to be taken online with no recall under incomplete knowledge, a decision tree should be paralleled by a probability model as a best choice model or online selection model algorithm. Another use of decision tree is as a descriptive means for calculating conditional probabilities.

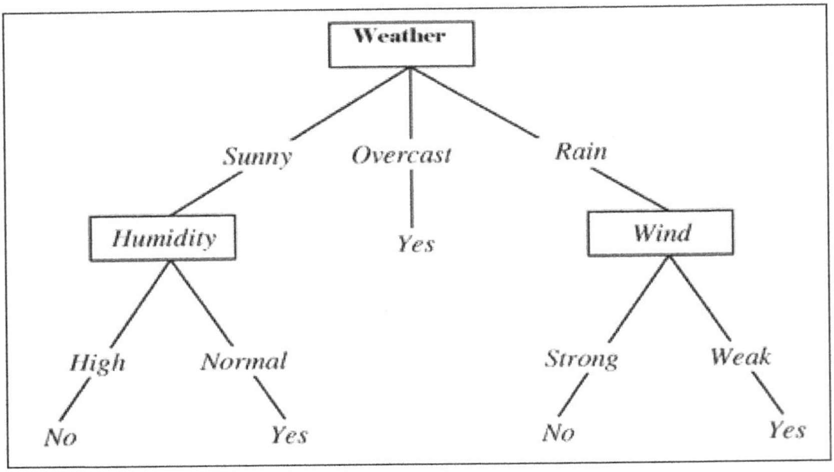

Figure 3.2.6.1: Decision Tree Example for Weather Forecasting

Decision tree are most powerful tool used for descriptive and predictive data modelling This can be used as a different approach to build a classification model. Due to '*represent rules*', the decision tree is more likely used techniques for key value-based searching. The DT produces a model of tree-shaped structure using inductive reasoning. To classify all input data, each node of the tree is a distinguish by an equation. Every equation will focus on a certain variable and determine whether the imported data is greater than, equal to, or less than a certain value. Each node of such can classify the imported data into different category afterwards. Decision tree is a common method that provides both classification and predictive functions simultaneously. With a sequence of questions and rules, data is classified (Chien et al., 2005). Featuring the use of similar models to predict similar outcomes, the decision tree theory is very suitable for conducting medical predictions and data analysis explanations in the field. Major algorithms of decision tree analysis model include ID3, C4.5, classification and regression trees (CRAT), chi-squared automatic interactive detector (CHAID) models and C5.0 algorithm. C5.0 can be most effectively used in processing huge data set specifically. Since it uses Boosting method to increase modelling accuracy, it is also known as Boosting Trees. It is a lot faster in speed and is more memory efficient than C4.5 (Chang and Chen, 2008).

Another important reason of using Decision tree is use of represent rules can be readily being expressed in natural language so that we human can understand them, and also can be expressed to database languages like SQL to retrieve records matching a particular criterion. Decision trees combine both data exploration and modelling, and so they are powerful first step in modelling process even when building a final model using another technique. In some application, the accuracy of a classification or prediction is most important things that matters for accuracy trade-off of the model. When decision trees are used for producing scores, having a large number of leaves is advantageous because each leaf generates a different score. When the objective is to create the rules, the fewer rules are better. And it is always possible to collapse a complex tree into smaller set of rules.

A decision tree represents series of questions, and the answer of the first question determines the follow-up question and so on. The initial question creates broad categories with many members; follow-on question divides the board categories into smaller and further smaller sets. If the questions are chosen properly, a surprisingly short series is enough to accurately classify an incoming record. The set of rules used in decision tree divides large heterogeneous population of data-set in to smaller and more homogeneous groups with respect to a target variable. In this way the tree describes a data-set containing half responders and half non-responders that can be also used to develop a response model.

Advantages of Decision Tree

- DT are simple to understand and interpret. People are able to understand decision tree models after a brief explanation.
- Have value even with little hard data. Important insights can be generated based on experts describing a situation (its alternatives, probabilities, and costs) and their preferences for outcomes.
- Possible scenarios can be added
- Worst, best and expected values can be determined for different scenarios
- DT uses a white box model. If a given result is provided by a model.

3.2.6.2 Artificial Neural Networks

The neural network is a popular technique used to track record in most of the data mining and decision support applications. It is a powerful, general purpose tool applied to prediction, classification and clustering. The basic element of a neural network is a neuron. This is a simple virtual device that accepts many inputs, sums them, applies a (usually nonlinear) transfer function, and generates the result, either as a model prediction or as input to other neurons. The term neural network was traditionally used to refer to a network or circuit of biological neurons (J. J. HOPFIELD 1982).

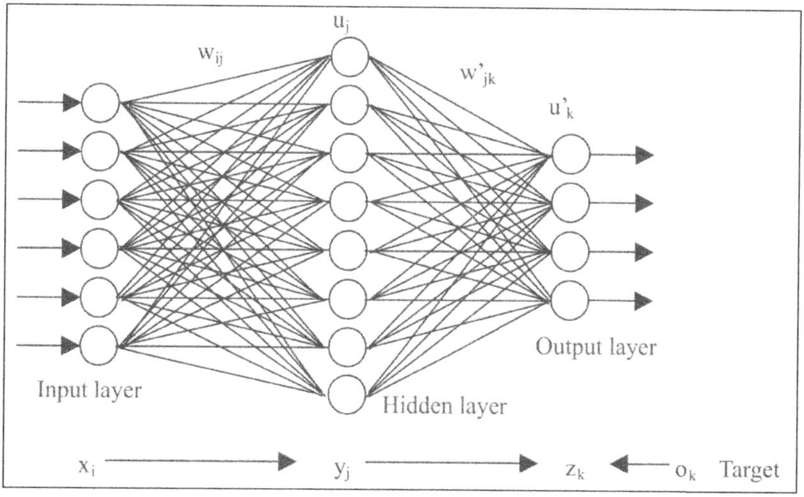

Figure 3.2.6.2-A Neural Network Example

The modern usage of the term often refers to artificial neural networks, which are composed of artificial neurons or nodes. Thus, the term has two distinct usages:

1. Biological neural networks are made up of real biological neurons that are connected or functionally related in a nervous system. In the field of neuroscience, they are often identified as groups of neurons that perform a specific physiological function in laboratory analysis.
2. Artificial neural networks are composed of interconnecting artificial neurons (programming constructs that mimic the properties of biological neurons) for solving artificial intelligence problems without creating a model of a real system. Neural network algorithms abstract away the biological complexity by focusing on the most important information. The goal of artificial neural networks is good, or human-like, predictive ability.

A biological neural network is composed of a group or groups of chemically connected or functionally associated neurons. A single neuron may be connected to many other neurons and the total number of neurons and connections in a network may be extensive. Apart from the electrical signalling, there are other forms of signalling that arise from neurotransmitter diffusion. Artificial intelligence and cognitive modelling try to simulate some properties of biological neural networks. While similar in their techniques, the former has the aim of solving particular tasks, while the latter aims to build mathematical models of biological neural systems. Neural networks are typically organized in layers. Layers are made up of a number of interconnected 'nodes' which contain an 'activation function'. Patterns are presented to the network via the input layer, which communicates to one or more 'hidden layers' where the actual processing is done via a system of weighted connections. The hidden layers then link of an output layer where the answer is output.

In the artificial intelligence field, artificial neural networks have been applied successfully to speech recognition, image analysis and adaptive control, in order to construct software. Most of the currently employed artificial neural networks for artificial intelligence are based on statistical, classification optimization and control theory. The cognitive modelling field involves the physical or mathematical modelling of the behaviour of neural systems; ranging from the individual neural level (e.g., modelling the spike response curves of neurons to a stimulus), through the neural cluster level (e.g., modelling the release and effects of dopamine in the basal ganglia) to the complete organism (e.g., behavioural modelling of the organism's response to stimuli). Artificial intelligence, cognitive modelling, and neural networks are information processing paradigms inspired by the way biological neural systems process data.

The rediscovery of the back-propagation algorithm was probably the main reason behind the re-popularization of neural networks after the publication of "Learning Internal Representations by Error Propagation" in 1986 (Though back-propagation itself dates from 1969). The original network utilized multiple layers of weight-sum units of the type.

$$f = g(w'x + b)$$

Where *g* was a sigmoid function or logistic function such as used in logistic - regression? Training was done by a form of stochastic gradient descent. The employment of the chain rule of differentiation in deriving the appropriate parameter updates results in an algorithm that seems to 'back-propagate errors', hence the nomenclature. However, it is essentially a form of gradient descent. Determining the optimal parameters in a model of this type is not trivial, and local numerical optimization methods such as gradient descent can be sensitive to initialization because of the presence of local minima of the training criterion. In recent times, networks with the same architecture as the back-propagation network are referred to as multilayer per captions. This name does not impose any limitations on the type of algorithm used for learning. The back -propagation network generated much enthusiasm at the time and there was much controversy about whether such learning could be implemented in the brain or not, partly because a mechanism for reverse signalling was not obvious at the time, but most importantly because there was no plausible source for the 'teaching' or 'target' signal. However, since 2006, several unsupervised learning procedures have been proposed for neural networks with one or more layers, using so-called deep learning algorithms. These algorithms can be used to learn intermediate representations, with or without a target signal, that capture the salient features of the distribution of sensory signals arriving at each layer of the neural network.

The working of neural network is similar of human brain that makes it possible for human to generalize from experiences. Computers, on the other hand, usually excel in following explicit instruction over and over. The appeal of neural network is that they fill this gap by modelling, on a digital computer. When we use neural network in well define domain, they have capability to generalize and learn mimics and to learn from experiences. A neural network is a structure of many such neurons connected in a systematic way. In Clementine, the neural networks used are feed-forward neural networks, also known as multilayer perceptions. The neurons in such networks (sometimes called units) are arranged in layers. Typically, there is one layer for input neurons (the input layer), one or more layers of internal processing units (the hidden layers), and one layer for output neurons (the output layer). Each layer is fully interconnected to the preceding layer and the following layer. For example, in a network with an input layer, a single hidden layer, and an output layer, each neuron in the input layer is connected to every neuron in the hidden layer, and each neuron in the hidden layer is connected to every neuron in the output layer. The connections between neurons have weights associated with them, which determine the strength of influence one neuron has on another. Information flows from the input layer through the processing layer(s) to the output layer to generate predictions. By adjusting the connection weights during training to match predictions to target values for specific records, the network "learns" to generate better and better predictions.

The neural network is good for prediction and estimation problems. And hence can be significantly useful with the problems where inputs of the problem are well understood, and you must have good idea of which feature of data are important, but you not necessarily know how to combine them. Another type of problem where the input is well understood and it is known that what is trying to be model.

Advantages of Neural Network

There are many advantages and limitations to neural network analysis and to discuss this subject properly we would have to look at each individual type of network, which isn't necessary for this general discussion. In reference to backpropagation networks however, there are some specific issues potential users should be aware of.

- Backpropagation neural networks (and many other types of networks) are in a sense the ultimate 'black boxes'. Apart from defining the general architecture of a network and perhaps initially seeding it with a random number, the user has no other role than to feed it input and watch it train and await the output. In fact, it has been said that with backpropagation, "you almost don't know what you're doing". Some software freely available software packages (NevProp, bp, Mactivation) do allow the user to sample the networks' progress at regular time intervals, but the learning itself progresses on its own. The final product of this activity is a trained network that provides no equations or coefficients defining a relationship (as in regression) beyond its own internal mathematics. The network 'IS' the final equation of the relationship.

- Backpropagational networks also tend to be slower to train than other types of networks and sometimes require thousands of epochs. If run on a truly parallel computer system this issue is not really a problem, but if the BPNN is being simulated on standard serial machine (i.e., a single SPARC, Mac or PC) training can take some time. This is because the machines CPU must compute the function of each node and connection separately, which can be problematic in very large networks with a large amount of data. However, the speed of most current machines is such that this is typically not much of an issue.

3.2.6.3 Sensitivity Analysis

Sensitivity analysis can be a manual process where each feature is tasted one at a time relative to other feature. It can also be automated process by using sensitive information generated by back propagation. Sensitivity analysis is an approach that analyses and reflects the sensitivity degree of how the outcome of the model can be apportioned and altered to different circumstances of variation (Chang and Chen, 2008). Generally, it first identifies the sensitive factors that affect the objects in the research. Then it confirms the degree of influence of a certain sensitive factor to the goal and objectives of the research. This approach can confirm variations and the changing trend of the goal and objectives in the research through obtaining variations regarding sensitive factors or information of the changing trend. Sensitivity analysis is performed to reduce network complexity by deleting the variables that have no or less influence on network training and to understand the degree of influence of each variable to network training. The greater the sensitivity degree, the larger the impact it has to the outcomes of artificial neural networks (Chang and Chen, 2008).

Sensitivity analysis can be used to get an idea of how opaque models work; sensitivity analysis does not provide explicit rules but it indicates the relative importance of the inputs to the result of the network. Sensitivity analysis uses the test set to determine how sensitive the output of the network at each input. The basic steps involve in sensitivity analysis is given below:

1. Find the average value for each input. We can think of this average value as the centre of the set.
2. Measure the output of the network when all inputs are at their average value.
3. Measure the output of the network when each input is modified, one at a time, to be at its minimum and maximum values.

For some inputs, the output of the network changes very little for the three values (minimum, average, and maximum). The network is not *sensitive* to these inputs at least when the inputs are the average values. Other inputs have a large effect on the output of the network and the network is sensitive to these inputs. The amount of change in the output measures the sensitivity of the network for each input.

3.3 Multi Agent Systems

Agents are Computer programs that use artificial intelligence technology to 'learn', and automate certain procedures and processes. For example, software agents at a newspaper website can learn about a user's preferences by tracking the user's actions, and then custom tailor the news summaries that suit the needs of that user. Agents technology is used also to automate ordering process according to pre-determined inventory levels or to search for cheaper sources of materials.

> *An agent is an encapsulated computer system that is situated in some environment and that is capable of flexible, autonomous action in that environment in order to meet its design objectives. A standard classification of agents is via the weak and strong notion of agency.* *(Jennings,2003).*

Agents are autonomous, computational entities that can be viewed as perceiving their environment through sensors and acting upon their environment through effectors. Agents are computational entities simply means that they physically exist in the form of programs that run on computing devices. Agents are autonomous means that to some extent they have control over their behaviour and can act without the intervention of humans and other systems. Agents pursue goals or carry out tasks in order to meet their design objectives, and in general these goals and tasks can be supplementary as well as conflicting.

Intelligent: Indicates that the agents pursue their goals and execute their tasks such that they optimize some given performance measures. To say that agents are intelligent does not mean that they are omniscient or omnipotent, nor does it mean that they never fail. Rather, it means that they operate flexibly and rationally in a variety of environmental circumstances, given the information they have and their perceptual and effectual capabilities.

Interacting: Indicates that the agents may be affected by other agents or perhaps by humans in pursuing their goals and executing their tasks. Interaction can take place indirectly through the environment in which they are embedded (e.g., by observing one another or by carrying out an action that modifies the environmental state) or directly through a shared language (e.g., by providing information in which other agents are interested or which confuses other agents).

It primarily focuses on coordination as a form of interaction that is particularly important with respect to goal attainment and task completion. The purpose of coordination is to achieve or avoid states of affairs that are considered as desirable or undesirable by one or several agents. To coordinate their goals and tasks, agents have to explicitly take dependencies among their activities into consideration. Two basic, contrasting patterns of coordination are cooperation and competition. In the case of cooperation, several agents work together and draw on the broad collection of their knowledge and capabilities to achieve a common goal. Cooperating agents try to accomplish as a team what the individuals cannot, and so fail or succeed together. Against that, in the case of competition, several agents work against each other because their goals are conflicting. Competitive agents try to maximize their own benefit at the expense of others, and so the success of one implies the failure of others.

An agent is a software program that acts flexibly on behalf of its owner to achieve particular objectives (Jennings,2003) To do this, the software must exhibit the following properties:

Autonomy: Capable of making decisions about what actions to take without constantly referring back to its user; i.e., to operate without direct intervention of humans, it operates without the direct intervention of humans or others and has control over its actions and internal state.

1. **Social Ability:** Ability to interact with other agents, i.e. It cooperates with humans or other agents in order to achieve its tasks.

2. **Reactive:** Able to respond appropriately to the prevailing circumstances in dynamic and unpredictable environments; i.e., perceive and respond to its environment. It perceives its environment and responds in a timely fashion to changes that occur in the environment.

3. **Proactive:** Able to act in anticipation of future goals so that its owner's objectives are met. i.e., take initiative. It does not simply act in response to its environment but is able to exhibit goal-directed behaviour by taking initiative.

An agent can be mobile, with the ability to travel between different nodes in a computer network. It can be rational, always acting in order to achieve its goals and never to prevent its goals being achieved, and it can learn, adapting itself to fit its environment and to the desires of its users.

3.3.1 Types of Agents

Haag (2006) suggests that there are four essential types of intelligent software agents:

1. **Buyer Agents or shopping bots**
2. **User or Personal Agents**
3. **Monitoring-and-Surveillance Agents**
4. **Data-Mining Agents**

Buyer Agents: Buyer agents travel around a network (i.e., the internet) retrieving information about goods and services. These agents, also known as 'shopping bots', work very efficiently for commodity products such as CDs, books, electronic components, and other one-size-fits-all products.

User Agents: User agents, or personal agents, are intelligent agents that take action on your behalf. In this category belong those intelligent agents that already perform, or will shortly perform, the following tasks:

- Check your e-mail, sort it according to the user's order of preference, and alert you when important emails arrive.
- Play computer games as your opponent or patrol game areas for you.
- Assemble customized news reports for you. There are several versions of these, including CNN.
- Find information for you on the subject of your choice.
- Fill out forms on the Web automatically for you, storing your information for future reference
- Scan Web pages looking for and highlighting text that constitutes the "important" part of the information there
- Facilitate with online job search duties by scanning known job boards and sending the resume to opportunities who meet the desired criteria
- Profile synchronization across heterogeneous social networks

Monitoring-and-Surveillance (Predictive) Agents: Monitoring and Surveillance Agents are used to observe and report on equipment, usually computer systems. The agents may keep track of company inventory levels, observe competitors' prices and relay them back to the company, watch stock manipulation by insider trading and rumours, etc. A special case of Monitoring-and-Surveillance agents are organizations of agents used to emulate the Human Decision-Making process during tactical operations. The agents monitor the status of assets (ammunition, weapons available, platforms for transport, etc.) and receive Goals (Missions) from higher level agents.

The Agents then pursue the Goals with the Assets at hand, minimizing expenditure of the Assets while maximizing Goal Attainment. (Popplewell, "Agents and Applicability")

Data Mining Agents: This agent uses information technology to find trends and patterns in an abundance of information from many different sources. The user can sort through this information in order to find whatever information they are seeking.

A data mining agent operates in a data warehouse discovering information. A 'data warehouse' brings together information from lots of different sources. "Data mining" is the process of looking through the data warehouse to find information that you can use to take action, such as ways to increase sales or keep customers who are considering defecting.

'Classification' is one of the most common types of data mining, which finds patterns in information and categorizes them into different classes. Data mining agents can also detect major shifts in trends or a key indicator and can detect the presence of new information and alert you to it. For example, the agent may detect a decline in the construction industry for an economy; based on this relayed information construction companies will be able to make intelligent decisions regarding the hiring/ firing of employees or the purchase/lease of equipment in order to best suit their firm.

3.3.2 Multi Agent Systems

Distributed artificial intelligence is a subfield of AI which is concerned with systems that consist of multiple independent entities that interact in a domain. Traditionally, this field has been broken into two sub-disciplines: distributed problem solving (DPS) and multiagent systems (MASs) (Stone and

Veloso, 2000). DPS focuses on developing solutions using the collective effort of multiple agents by combining their knowledge, information, and capabilities. It is often applied to solve large problems by decomposing them into smaller subtasks, each assigned to a different agent, and can therefore be regarded as a tightly-coupled top-down approach. A MAS, on the other hand, is a loosely-coupled bottom-up approach that aims to provide principles for the construction of complex systems consisting of multiple independent agents, and focuses on the coordination of the behaviours of the agents in such systems (Stone and Veloso, 2000). The agents in a MAS can have different, even conflicting, goals. Collaboration enables the different entities to work more efficiently and to complete activities they are not able to accomplish individually. More specifically, the use of a MAS has the following advantages (Sycara, 1998; Stone, 1998; Vlassis, 2003):

The existence of multiple agents can speed up the operation of a system because the agents can perform the computations in parallel. This is especially the case for domains in which the overall task can be decomposed into several independent subtasks that can be handled by separate agents.

- A MAS usually has a high degree of robustness. In single-agent systems a single failure can cause the entire system to crash. A MAS on the other hand degrades gracefully: if one or several agents fail, the system will still be operational because the remaining agents can take over the workload.
- The modularity of MAS enables one to add new agents to the system when necessary, and therefore has a high scalability. Adding new functionality to a monolithic system is often much more difficult.

- A MAS can make observations and perform actions at multiple locations simultaneously, and therefore it can take advantage of geographical distribution.
- A MAS usually has a higher performance-cost ratio than single-agent systems. A single robot with all the necessary capabilities for accomplishing a task is often much more expensive than the use of multiple, cheaper, robots each having a subset of these capabilities.

3.3.3 Cognitive Parameters

The following parameters are termed as cognitive. Their definitions are given below.

Belief: The belief corresponds to information the agent has about the world, desires correspond to the world state the agent wishes for, and intentions are the desires that the agent has selected to pursue. Belief is considered as an informational attitude. Viewed from the temporal perspective an agent's beliefs originate from a history of experiences; for example, observations and received communications. Moreover, beliefs affect future actions of the agent via their impact on other intentional notions (e.g., intentions or desires) that form the basis of actions. Beliefs are related to history of agents. Their relation to the future is addressed in the temporal description of the other, motivational attitudes (Bell, 1995).

Desire: Desires can refer to a (desired) state of affairs in the world (and the other agents), but also to (desired) actions to be performed (Frances, Barbara, Treur, Verbrugge, 1997). An agent's desires are conceived of as the states of the world that the agent wishes to bring about (Bell, 1995 Kraus, 1998). Desires represent the tasks allocated to the agent and so correspond to the objectives, or goals, it should accomplish.

Preference: An agent's preference plays an active role in social practical reasoning, where an action is to be selected in order for given intention to be fulfilled (Panzarasa, Jennings, Norman, 2002).

Intention: Intention is actually the goal that an agent wants to achieve; an agent's intention will give a guideline of what to do (Wu, Ekaette, Far, 2003). Intention is a fundamental characteristic that agents involve a special kind of "self-commitment" to acting (Bratman 1990, Von 1980).

Commitment: An agent's individual intention towards a state of affairs entails the agent's commitment to acting towards the achievement of that state (Panzarasa, Jennings, Norman, 2002).

Performance: The performance of an agent can be measured with respect to the task or tasks it is performing. Three types of performance measures are commonly employed: effectiveness (is the task being performed well), efficiency (is the task being performed in such as way that output is maximized relative to some input), and perceived effectiveness (is the agent perceived as performing well by one or more stakeholders such as the general public, the government, the board of directors, or the media) (Gerhard et al, 2001).

Capability: Capability needed by an agent is the ability to decide on its own actions in the context of the activities of other agents around it (Gerhard et al, 2001).

Trust: Trust is commonly used parameters/features that depict the performance of agent in a business domain. There are many definitions of trust depending upon the problem domain. Some of the important and pertaining definitions by different researchers are given below:

Trust is a set of expectations that lead to behavioural intentions that involve potential loss, because of the absence of control over those upon whom one depends. Trust is defined as "reliance upon the characteristics of an object, or the occurrence of an event, or the behaviour of a person in order to achieve a desired but uncertain objective in a risky situation" (Giffin, 2003). This trust typically consists of the three trust dimensions: intelligence (corresponding to ability), good character (honesty and integrity), and good will (benevolence).

Rotter also defined trust as "an expectancy held by an individual or a group that the word, promise, verbal or written statement of another individual or group can be relied on". Again, trust is defined as a situation-specific expectation or belief that determines behaviour.

Zucker has defined trust as a set of shared social expectations that are essential for and determine social behaviour, enabling individuals to respond to each other without the explicit specification of contractual details. Similarly, Korsgaard et al. argue that trust is the confidence that team members have in the good will and honesty (that is to say, benevolence and integrity) of their leader, while Hart and Saunders have defined trust as the confidence that another party will behave as expected, combined with expectations of the other party's good will. This trust is composed of a perception of the partners' competence, openness (the willingness not to withhold information), caring (not taking unfair advantage), and reliability.

McAllister defines trust as "the extent to which a person is confident in, and willing to act on the basis of, the words, actions, and decisions of another", that is, as a behavioural intention based on beliefs.

Kumar and Ramaswami et al. define trust as dependability based upon perceptions of procedural justice – fairness in managing a relationship, policies and procedures – and distributive justice – the perceived fairness of outcomes.

Rousseau et al. defined that trust deals with "the intention to accept vulnerability based upon positive expectations of the intentions or behaviour of another". Consistent with Lewis and Weigert, Rousseau et al. also note that trust differs depending upon the history and nature of the relationship between the parties.

Dwyer et al. define trust as a set of beliefs relating to the exchange partner's ability and willingness to take part in the social exchange. Moorman et al. define trust is "a willingness to rely on an exchange partner in whom one has confidence". Morgan and Hunt regard trust as confidence in another person's reliability and integrity. Ganesan also defines trust as a willingness to depend upon another, based on beliefs or expectation resulting from the partner's experience, reliability and benevolence. Gefen (2002), in the context of e-Commerce, defines trust as a single dimension construct dealing with a consumer's assessment that the vendor is trustworthy, based on Luhmann's (1995) definition of trust as a social complexity-reducing mechanism that leads to a willingness to depend on a vendor; this willingness is derived from the perception that the vendor will fulfil its commitments.

Reputation: Reputation is considered on the basis of two users rating (Zacharia, 1999; Sabater, Sierra, 2005). Histos (Zacharia, 1999; Sabater, Sierra, 2005) reputation definition is based upon the user witness information i.e., the most recent experience of the broker and third-party agent with the agent that is being evaluated. Therefore, the reputation value of each user is a global measure shared by all the observers.

3.3.4 Multi Agent System Communication

Agents are fundamentally a form of distributed code processes and thus comply with the classic notion of a distributed computing model comprising two parts: components and connectors. Components are consumers, producers and mediators of communication messages exchanged via connectors.

Foundation for Intelligent Physical Agents (FIPA): The Foundation for Intelligent Physical Agents (FIPA) is an international non-profit association of companies and organizations sharing the effort to produce specifications of generic agent technologies. FIPA is envisaged not just as a technology for one application but as generic technologies for different application areas, it is a set of basic technologies that can be integrated by developers to make complex systems with a high degree of interoperability. At the core of FIPA is the following set of principles:

1. Agent technologies provide a new paradigm to solve old and new problems.
2. Some agent technologies have reached a considerable degree of maturity.
3. To be of use some agent technologies require standardization.
4. Standardization of generic technologies has been shown to be possible and to provide effective results by other standardization for an agent.
5. The standardization of the internal mechanics of agents themselves is not the primary concern, but rather the infrastructure and language required for open interoperation.

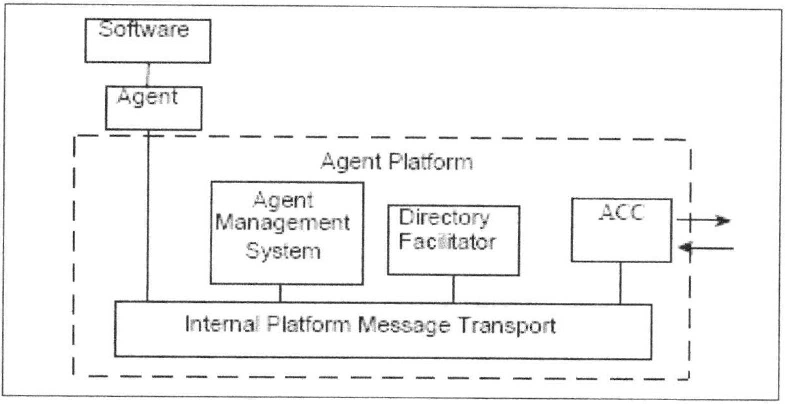

Figure: 3.3.4- FIPA reference model

In FIPA the agent communication is based on message passing, where agents communicate by formulating and sending individual messages to each other. The FIPA ACL specifies a standard message language by setting out the encoding, semantics and pragmatics of the messages. The standard does not set out a specific mechanism for the internal transportation of messages. Instead, since different agents might run on different platforms and use different networking technologies, FIPA specifies that the messages transported between platforms should be encoded in a textual form. It is assumed that the agent has some means of transmitting this textual form. The syntax of the ACL is very close to the widely used communication language KQML. However, despite syntactic similarity, there are fundamental differences between KQML and ACL, the most evident being the existence of a formal semantics for ACL which should eliminate any ambiguity and confusion from the usage of the language. The standard supports common forms of inter-agent conversations through the specification of interaction protocols, which are patterns of messages exchanged by two or more agents. Such protocols include range from simple query-request protocols, to the well-known contract net negotiation protocol and English and Dutch auctions. Other parts of the FIPA standard specify other aspects, in particular the agent-software integration, agent mobility and security, ontology service, and the Human-Agent Communication. However, they have not yet been considered into the JADE implementation.

FIPA-compliant Agent Platform, which includes the AMS (Agent Management System), the DF (Directory Facilitator), and the ACC (Agent Communication Channel). All these three agents are automatically activated at the agent platform start-up. The agent platform can be split on several hosts (provided that there is no firewall between them). A number of FIPA-compliant DFs (Directory Facilitator) can be started at run time in order to implement multi-domain applications.

Agent communication is implemented using FIPA ACL messages (FIPA, 1999). FIPA used the following messages performatives as given in table 2.2. SUBSCRIBE messages are used by the seller and buyer agents. INFORM messages are used as responses to SUBSCRIBE or REQUEST messages. Buyer agents are using FAILURE messages to inform the broker or seller agents about the unsuccessful result of a negotiation. Finally, CFP, PROPOSE, ACCEPT-PROPOSAL, REJECT-PROPOSAL and REFUSE messages are being used by negotiating agents. (Ghanza, et al 2002). The various FIPA communication acts (performatives) are given below in table.

Table: 3.3.4 - FIPA Communication Acts

The FIPA communicative acts	
FIPA communicative act	**Description**
Accept Proposal	The action of accepting a previously submitted proposal to perform an action
Agree	The action of agreeing to perform some action, possibly in the future

Cancel	The action of one agent informing another agent that the first agent no longer has the intention that the second agent performs some action
Call for Proposal	The action of calling for proposals to perform a given action
Confirm	The sender informs the receiver that a given proposition is true, where the receiver is known to be uncertain about the proposition.
Disconfirm	The sender informs the receiver that a given proposition is false, where the receiver is known to believe, or believe it likely that, the proposition is true.
Failure	The action of telling another agent that an action was attempted but the attempt failed.
Inform The sender informs the receiver that a given proposition is true.	
Inform If	A macro action for the agent of the action to inform the recipient whether or not a proposition is true.
Inform Ref	A macro action allowing the sender to inform the receiver of some object believed by the sender to correspond to a specific descriptor, for example a name.
Not Understood	The sender of the act (for example, i) informs the receiver (for example, j) that it perceived that j performed some action, but that i did not understand what j just did. A particular common case is that i tells j that i did not understand the message that j has just sent to i.
Propagate	The sender intends that the receiver treat the embedded message as sent directly to the receiver, and wants the receiver to identify the agents denoted by the given descriptor and send the received propagate message to them
Propose	The action of submitting a proposal to perform a certain action, given certain preconditions
Proxy	The sender wants the receiver to select target agents denoted by a given description and to send an embedded message to them
Query If	The action of asking another agent whether or not a given proposition is true
Query Ref	The action of asking another agent for the object referred to by a referential expression

Refuse	The action of refusing to perform a given action, and explaining the reason for the refusal
Reject Proposal	The action of rejecting a proposal to perform some action during a Negotiation
Request	The sender requests the receiver to perform some action One important class of uses of the request act is to request the receiver to perform another communicative act
Request When	The sender wants the receiver to perform some action when some given. Proposition becomes true
Request Whenever	The sender wants the receiver to perform some action as soon as some proposition becomes true and thereafter each time the proposition becomes true again
Subscribe	The act of requesting a persistent intention to notify the sender of the value of a reference, and to notify again whenever the object identified by the reference changes

3.3.5 MAS Software (JADE)

JADE (Java Agent Development Framework) is a software framework to make easier the development of agent applications in compliance with the FIPA specifications for interoperable intelligent multi-agent systems. The goal of JADE is to simplify development while ensuring standard compliance through a comprehensive set of system services and agents.

To achieve such a goal, JADE offers the following list of features to the agent programmer:

Architecture/Function: JADE fully complies with the reference architecture of a FIPA agent platform. JADE architecture has an Agent Management System (AMS) which has supervisory control over agent platform execution. AMS has a directory of agent identifiers and the state of agent AMS also provide life cycle of an agent. The Agent Communication Channel (ACC) is the software component which controls the exchange of messages with in the platform including remote platform i.e., part of this architecture. JADE is an enabling technology, a middle ware for the development and run time execution of peer-to-peer applications which are based on the agent's paradigm. JADE works as an operating system for the agent-based system. JADE is composed with the main functionalist (packages). To implement the systems, kernel the package is JADE core. It has sub package which contain behaviour class hierarchy for implement the task on intentions of agent. JADE contains different packages for communication, massage transport, display and edit etc.

The Middleware: The term middleware is meant to describe all those higher-level libraries that enable easier and more effective application development by providing generic services, variety of applications and controls.

Tools available in JADE: RMA, Dummy Agent, Sniffer, Log Manager Agent, Introspected are available tools in JADE are used for platform management and control, monitoring and debugging, compose and manage ACL messages, monitoring the life cycle of an agent etc.

Implementation: JADE is a software development frame work aimed at developing multi agent system and application conforming to FIPA standards for intelligent agent. Mainly it provides FIPA compliant agent platform and a back age to develop Java agents. JADE is written in JAVA language using various Java packages. It is an object-oriented programming in distributed heterogeneous environments with following main features: object serialization, reflection API and Remote Method Invocation (RMI). JADE is composed of the various packages like jade.core for implement the kernel; jade.lang.all for agent communication, jade.gui for display etc.

Interface: JADE provides interface for custom application dependent task. In this way the agent platform can be split in several host and they connected via RMI. The debugging tools help in developing the multivalent applications is based on JADE.

FIPA compliant agent platform automatically activated at the JADE agent platform. It supports intra-platform agent mobility including the transfer of both the state and the code of agent. Hence JADE is preferable and useful to the users. It simplifies the distributed applications composed of autonomous entities and the development of application that requires negotiation and coordination among a set of agents. The agents control their own thread of execution without any human intervention. It is an open source and its API'S are easy to learn and use.

3.3.6 Agent Designing Issues

Issues to consider in the development of agent-based systems includes:

- How tasks are scheduled and how synchronization of tasks is achieved?
- How tasks are prioritized by agents?
- How agents can collaborate, or recruit resources?
- How agents can be re-instantiated in different environments, and how their internal state can be stored?
- How the environment will be probed and how a change of environment leads to behavioural changes of the agents?
- How messaging and communication can be achieved?
- What hierarchies of agents are useful? (e.g., task execution agents, scheduling agents, resource providers).

For software agents to work together efficiently they must share semantics of their data elements. This can be done by having computer systems publish their metadata.

The definition of agent processing can be approached from two interrelated directions:

- Internal state processing and ontologies for representing knowledge.
- Interaction protocols - standards for specifying communication of tasks.

Agent systems are used to model real-world systems with concurrency or parallel processing.

- **Agent Machinery** - Engines of various kinds, which support the varying degrees of intelligence.
- **Agent Content** - Data employed by the machinery in Reasoning and Learning.
- **Agent Access** - Methods to enable the machinery to perceive content and perform actions as outcomes of Reasoning.
- **Agent Security** - Concerns related to distributed computing, augmented by a few special concerns related to agents.

The agent uses its access methods to go out into local and remote databases to forage for content. These access methods may include setting up news stream delivery to the agent, or retrieval from bulletin boards, or using a spider to walk the Web.

The content that is retrieved in this way is probably already partially filtered – by the selection of the newsfeed or the databases that are searched. The agent next may use its detailed searching or language-processing machinery to extract keywords or signatures from the body of the content that has been received or retrieved.

This abstracted content (or event) is then passed to the agent's Reasoning or inferencing machinery in order to decide what to do with the new content. This process combines the event content with the rule-based or knowledge content provided by the user. If this process finds a good hit or match in the new content, the agent may use another piece of its machinery to do a more detailed search on the content. Finally, the agent may decide to take an action based on the new content; for example, to notify the user that an important event has occurred. This action is verified by a security function and then given the authority of the user. The agent makes use of a user-access method to deliver that message to the user. If the user confirms that the event is important by acting quickly on the notification, the agent may also employ its learning machinery to increase its weighting for this kind of event.

3.3.7 Methodology for Agent –Oriented Software Design

The potential of agent-based systems is large and it remains to exploit this new technology. The Success of agent-based systems in the future, in contrast, in our view depends on a thorough analysis of its conceptual basis and the construction of structuring principles and methods for the design of agent systems. Despite the many proposals in the literature to support the construction and design of agents by means of specific agent architectures or programming languages, there are still few methodologies for the development of multi-agent systems. Some promising efforts in providing a methodology for agent-oriented software engineering derived from object-oriented methodologies have been proposed. These proposals provide a first step towards an approach to software design based on agent-related concepts. Nevertheless, the additional complexity of inter-agent communication and coordination requires a major effort which involves the development of substantially new concepts which is not fully accomplished in the proposals so far. These new concepts are related to the high-level nature of the information present in multi-agent systems. The high-level nature of the information present in multi-agent systems which involve, for example, mechanisms for update and belief revision, in general introduces a completely new dimension to existing design methodologies.

The aim of the agent-oriented approach is to facilitate analysis, design and implementation, at a level of abstraction that is more adequate for complex information systems, by introducing the single, main concept of agent. We provide an outline of the basic ideas that we think are useful to this end and a preliminary sketch of an agent-oriented design methodology. Of course, it is the purpose of this proposal to refine, evaluate and test, and extend this sketch of a methodology. The sketch provided here serves to illustrate the main ideas and to introduce the main research issues.

The basis for a methodology for agent-oriented design can be derived from the most important concepts in the agent-oriented approach, namely that of agent and group. A third design issue - that of the interaction between agents and groups - is inferred immediately from these notions. One of the basic concepts in our proposal for a methodology is that of a role. Roles are associated with agents and are composed of skills, knowledge and goals. In our design methodology a number of steps concern the further refinement of these skills, knowledge and goals of roles.

Frank S. De Bore proposes that an agent-oriented methodology should consist of the construction of five main models:

Each of the steps in the methodology below results in the construction of one or more of the corresponding models. The methodology for their elaboration and refinement can be expressed in five steps.

1. The first step consists of the identification of the roles in the application domain in terms of the corresponding goals or tasks. Consequently, a role is identified by the main tasks or responsibilities required by the system and form the basis for a specification of agent types analogously to the concept of class in object-oriented programming (OO). There is a close link here with similar concepts in OO. The research issues involve defining notions like inheritance, agent classes, aggregation, etc. in the agent context.

2. Secondly, identify the main skills associated with agent roles - skills consist of the responsibilities and basic services required to be able to perform a role. Special attention will be addressed to the issue of agent-wrapping: the integration of non-agent software components in this step, and the design of appropriate interfaces.

3. A third step consists of the modelling of the knowledge about the application domain associated with identified roles or skills and should result in the design of an adequate ontology. Techniques from knowledge engineering can be used here, and again the issue of reuse and integration of existing technologies will be addressed.

4. Fourthly, an organisational structure (architecture) of the multi-agent system should be designed. The coordination between agents to achieve a task and the required communication schemes need to be analysed in

this step, and group structures need to be identified. Groups allow sets of agents to be treated as a single conceptual entity. The main research themes related to this step concern the design of generic group structures, communication schemes and the overall infrastructure of the multi-agent system.

5. Finally, the dynamics of the multi-agent system should be analysed in terms of the flow of information, resulting in, for example, and synchronisation requirements derived from the roles associated with individual agents.

The methodology outlined above is grounded in the agent programming framework 3APL which facilitates both the internal design of agents as well as the implementation of the multi-agent system. 3APL grounds the methodology in providing clear interpretations of the basic notions of belief and goals and plans of an agent. Finally, the communication primitives of 3APL provide for high-level information exchange capabilities of agents which are suitable for refining and implementing the communication schemes of step 4. Apart from the analysis steps 1 and 5, 3APL provides thus a concrete platform which substantially reduces the effort needed to refine three of the five types of models which are constructed during the design of a software system. We are currently implementing a programming environment for 3APL. This implementation provides a test-bed for the evaluation of a methodology for the design of agent systems and the construction of an agent application. The various themes which we believe should play a major role in an agent-oriented methodology and elaborate on some of the associated issues which require further investigation. The themes correspond with issues which need to be addressed in the first four steps of the methodology.

3.3.7.1 Agent Roles

In elaborating and refining the methodology for agent-oriented design, we aim at a further development of the concept of a class from object-oriented programming, which we call a role. In contrast to a class, however, roles offer a more goal or action-oriented perspective. A role should specify at an appropriate abstraction level the set of possible behaviours of an agent in terms of the services associated with it and its high-level interaction capabilities. Roles can be organisational - in case they are required by system implementation - or functional - in case they are directly related to the application. The concept of a role is expected to provide a basis for the description and development of generic agent architectures. Typical examples include agent classes like brokers, mediators, and facilitators used for delivering information to the agents with the appropriate processing skills, and contracting agents to perform services associated with their roles. As abstract specification formalism for roles, we envisage a graphical notation, like for example UML, with a declarative interpretation.

3.3.7.2 Agent Wrapping

Another important design issue concerns the integration of existing software, like existing database applications, and software techniques into agent-based systems. Agent-based systems constructed in 3APL are not introduced as a new and self-contained way of building software. In contrast, 3APL provides a programming environment that explicitly aims at the exploitation and reuse of existing software. This aim, however, poses new challenges and design choices as to how to integrate a wide range of heterogeneous system components into the basic agent design and finally how to refine this design into the basic structure offered by 3APL.

The choices involve the integration of existing information sources like databases and the Web, the sharing, manipulation and accessing of information between agents, and issues like security and privacy.

3.3.7.3 Knowledge Modelling

First of all, the focus in the agent-oriented approach is on the processing of information and therefore it is of vital importance that the beliefs of an agent are adequately modelled when designing agent-based systems. More specifically, the combination of issues like persistence, and the reliability of other agents introduce a very different and new picture.

A characteristic of 3APL is that it does not fix the type of information that is processed or exchanged between agents and as a consequence does not impose many restrictions (as some existing agent architectures do) but allows for a wide range of information agents. As such, it provides a generic mechanism for turning a database into an agent, that is, a generic mechanism for agent wrapping of databases but at the same time raising the important issue of designing appropriate ontologies and associated problem-solving methods.

3.3.7.4 Coordination and Groups

Agent-based systems provide for different levels of abstraction which are built into the very notion of a multi-agent system. This provides for a potentially powerful and new design approach. It also poses the problem of structuring and articulating the different levels in such a way that they are most useful for the design of the type of information systems considered here. Apart from the design level of the individual agent, a no less important level is that of the multi-agent system itself. This level concerns the interoperability of agents. The notion of a group here complements that of a role. Similar issues like finding a balance between private and public information reappear and new issues like group-structure, and hierarchy relations between agents need to be investigated. In particular, the strategies or protocols for information exchange and negotiation between agents, within and outside a group, provide new challenges. A major challenge therefore consists in the development of design patterns for the distribution of databases among agents.

3.3.7.5 Programming Environment and Application

At a more practical level the research described above should lead to the development of tools to support the design of agent-based systems in 3APL by means of a programming environment and the actual design and implementation of a substantial application. Related to this we would like to develop libraries consisting of basic agent skills and generic agent types or architectures, like, for example, a broker agent. A particularly interesting application, both from a practical as well as from a more scientific perspective, is so-called personal assistants. These types of agents involve activities like maintaining an agenda of their user, actively filtering and suggesting relevant information sources, and negotiating with other agents.

3.4 Geographic Information System

Geospatial information lends itself to many opportunities that can result in better decisions, higher productivity and enhanced efficiencies in agriculture. Here is a look at how various geospatial technologies can be applied to agricultural processes and the benefits accrued. Geospatial is one such technology. One reason why this technology has a significant role to play is because of the spatial nature of agriculture itself.

Agriculture and natural resources are essentially understood by a geographic location.

GIS can store layers of information, such as yields, soil survey maps, crop scouting reports and soil nutrient levels. GIS can display geographically referenced data, adding a visual perspective for interpretation. In addition to data storage and display, GIS can be used to evaluate present and alternative management by combining and manipulating data layers to produce an analysis of management scenarios, informs Anil Kumar Singh, Indian Agricultural Research Institute.

GIS has a wide range of possible applications at the national and local level. According to M.V.K. Sivakumar and Donald E. Hinsman of Agricultural Meteorology Division and Satellite Activities Office, World Meteorological Organization, geographical data can assist agricultural planners in deciding on factors like the best zones for cash crop, combining data on soils, topography, and rainfall to determine the size and location of biologically suitable areas. It can also provide information on related aspects like land ownership, transport, infrastructure, labour availability, and distance to market centres. Elaborating on the role of GIS in agriculture, Matthew observes that GIS in agriculture is about allocative efficiency, profitability, and record keeping. Spatial thinking enables producers to minimise inputs and maximise outputs. GIS provides mechanisms to better understand environmental, policy, and market forces from local to global scales. In an increasingly regulatory environment, GIS offers a means to develop real time records of applications of fertilisers, herbicides, etc. and quantify biomass, increasing profitability.

Sivakumar and Donald inform that agricultural weather and climate data systems are needed to expedite generation of products, analyses and forecasts that affect agricultural cropping and management decisions, irrigation scheduling, commodity trading and markets, fire weather management and other preparedness for calamities, and ecosystem conservation and management. One of the most significant benefits of geospatial technology is that it touches agricultural processes at various stages and offers potential to meet requirements of various stakeholders at different levels, thereby addressing in the entire life cycle of agriculture.

Matthew opines that productivity, by definition, is primarily concerned with outputs. In that context, the GIS solutions offered by the company through ArcGIS helps enhance agricultural productivity on the farm by analysing field and crop conditions, monitoring environmental patterns such as weather, and developing input recommendations for the right place, at the right time, in the most efficient manner. Desktop, mobile, server and online components of GIS systems are now completely integrated. For agriculture, this means that the same platform can be used by producers and crop consultants for field-based management and analysis; agribusiness for customer and market development; and governments and associations for regional to global analysis and monitoring. (Geospatial World I August 2011)

GIS maps are interactive. On the computer screen, map users can scan a GIS map in any direction, zoom in or out, and change the nature of the information contained in the map. They can choose whether to see the roads, how many roads to see, and how roads should be depicted. Then they can select what other items they wish to view alongside these roads such as storm drains, gas lines, rare plants, or hospitals. Some GIS programs are designed to perform sophisticated calculations for tracking storms or predicting erosion patterns. GIS applications can be embedded into common activities such as verifying an address. Given a long association with Esri products and a significant investment by the spatial analysis.

3.4.1 GIS Techniques and Technology

Modern GIS technologies use digital information, for which various digitized data creation methods are used. The most common method of data creation is digitization, where a hard copy map or survey plan is transferred into a digital medium through the use of a CAD program, and geo-referencing capabilities. With the wide availability of ortho-rectified imagery (both from satellite and aerial sources), heads-up digitizing is becoming the main avenue through which geographic data is extracted. Heads-up digitizing involves the tracing of geographic data directly on top of the aerial imagery instead of by the traditional method of tracing the geographic form on a separate digitizing tablet.

3.4.1.1 Relating Information from Different Sources

GIS uses spatial-temporal location as the key index variable for all other information. Just as a relational database containing text or numbers can relate many different tables using common key index variables, GIS can relate unrelated information by using location as the key index variable. The key is the location and/or extent in space-time.

Any variable that can be located spatially, and increasingly also temporally, can be referenced using a GIS. These GIS coordinates may represent other quantified systems of temporo-spatial reference (for example, film frame number, stream gage station, highway mile-marker, surveyor benchmark, building address, street intersection, entrance gate, water depth sounding, POS or CAD drawing origin/units). Units applied to recorded temporal-spatial data can vary widely (even when using exactly the same data, see map projections), but all Earth-based spatial–temporal location and extent references should, ideally, be relatable to one another and ultimately to a "real" physical location or extent in space–time.

Related by accurate spatial information, an incredible variety of real-world and projected past or future data can be analysed, interpreted and represented to facilitate education and decision. (Cowen 1988). This key characteristic of GIS has begun to open new avenues of scientific inquiry into behaviours and patterns of previously considered unrelated real-world information.

3.4.1.2 GIS Uncertainties

GIS accuracy depends upon source data, and how it is encoded to be data referenced. Land surveyors have been able to provide a high level of positional accuracy utilizing the GPS-derived positions. High-resolution digital terrain and aerial imagery, powerful computers and Web technology are changing the quality, utility, and expectations of GIS to serve society on a grand scale, but nevertheless there are other source data that have an impact on overall GIS accuracy like paper maps, though these may be of limited use in achieving the desired accuracy since the aging of maps affects their dimensional stability.

In developing a digital topographic data base for a GIS, topographical maps are the main source of Aerial photography and satellite images are extra sources for collecting data and identifying attributes which can be mapped in layers over a location facsimile of scale. The scale of a map and geographical rendering area representation type are very important aspects since the information content depends mainly on the scale set and resulting locatability of the map's representations.

In order to digitize a map, the map has to be checked within theoretical dimensions, then scanned into a raster format, and resulting raster data has to be given a theoretical dimension by a rubber sheeting/warping technology process. A quantitative analysis of maps brings accuracy issues into focus. The electronic and other equipment used to make measurements for GIS is far more precise than the machines of conventional map analysis All geographical data are inherently inaccurate, and these inaccuracies will propagate through GIS operations in ways that are difficult to predict.

3.4.1.3 Data Representation

GIS data represents real objects (such as roads, land use, elevation, trees, waterways, etc.) with digital data determining the mix. Real objects can be divided into two abstractions: discrete objects (e.g., a house) and continuous fields (such as rainfall amount, or elevations). Traditionally, there are two broad methods used to store data in a GIS for both kinds of abstractions mapping references: raster images and vector. Points, lines, and polygons are the stuff of mapped location attribute references. A new hybrid method of storing data is that of identifying point clouds, which combine three-dimensional points with RGB information at each point, returning a "3D colour image". GIS thematic maps then are becoming more and more realistically visually descriptive of what they set out to show or determine.

3.4.1.4 Data Capture

Data capture is entering information into the system, consumes much of the time of GIS practitioners. There are a variety of methods used to enter data into a GIS where it is stored in a digital format. Existing data printed on paper or PET film maps can be digitized or scanned to produce digital data. A digitizer produces vector data as an operator traces points, lines, and polygon boundaries from a map.

3.4.1.5 Raster-to-Vector Translation

Data restructuring can be performed by a GIS to convert data into different formats. For example, a GIS may be used to convert a satellite image map to a vector structure by generating lines around all cells with the same classification, while determining the cell spatial relationships, such as adjacency or inclusion.

More advanced data processing can occur with image processing, a technique developed in the late 1960s by NASA and the private sector to provide contrast enhancement, false colour rendering and a variety of other techniques including use of two-dimensional Fourier transforms. Since digital data is collected and stored in various ways, the two data sources may not be entirely compatible. So, a GIS must be able to convert geographic data from one structure to another.

3.4.1.6 Projections, Coordinate Systems and Registration

The earth can be represented by various models, each of which may provide a different set of coordinates (e.g., latitude, longitude, elevation) for any given point on the Earth's surface. The simplest model is to assume the earth is a perfect sphere. As more measurements of the earth have accumulated, the models of the earth have become more sophisticated and more accurate. In fact, there are models called datums that apply to different areas of the earth to provide increased accuracy, like NAD27 for U.S. measurements, and the World Geodetic System for worldwide measurements.

3.4.1.7 Spatial Analysis with GIS

GIS spatial analysis is a rapidly changing field, and GIS packages are increasingly including analytical tools as standard built-in facilities, as optional toolsets, as add-ins or 'analysts'. In many instances these are provided by the original software suppliers whilst in other cases facilities have been developed and are provided by third parties. Furthermore, many products offer software development kits (SDKs), programming languages and language support, scripting facilities and/or special interfaces for developing one's own analytical tools or variants. The increased availability has created a new dimension to business intelligence termed "spatial intelligence" which, when openly delivered via intranet, democratizes access to geographic and social network data. GIS spatial analysis has also become a key element for security intelligence.

3.4.2 Precision Agriculture

Precision agriculture is an agricultural system that has the potential of dramatically changing agriculture in this 21st century. Precision agriculture lends itself to most agricultural applications and can be implemented at whatever levels are required. Precision agriculture is based on information technology, which enables the producer to collect information and data for better decision making. Precision agriculture is a pro-active approach that reduces some of the risk and variables common to agriculture. Precision agriculture is more environmentally sound and is integral part in sustaining natural resources. To better understand the need for an accurate definition of precision agriculture let's look at how precision agriculture is being considered. Precision agriculture is considered a concept, management strategy, and even a philosophy. It is said, "Precision agriculture is a phrase that captures the imagination of many concerned with the production of food, feed, and fiber." The concept of precision agriculture offers the promise of increasing productivity while decreasing production cost and minimizing environmental impacts. Precision agriculture conjures up images of farmers overcoming the elements with computerized machinery that is precisely controlled via satellites and local sensors and using planning software that accurately predicts crop development. This image has been called the future of agriculture.

Precision farming is characterized by a number of sophisticated tools that assist in monitoring variation and managing inputs. These include:

- Global Positioning System (GPS) - a referencing device capable of identifying sites within a field.
- Sensors and dataloggers - crop, soil and climate information can be monitored at a high frequency using these technologies.
- Geographic Information Systems (GIS) - maps of these attributes can be generated and analysed using simple browsers or complex models.

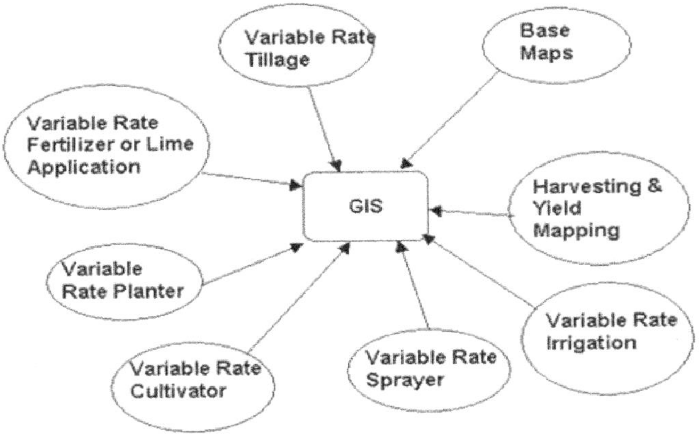

Figure: 3.4.2- Application of GIS for precision farming

Differential Action: Once the factors that are limiting yield are identified, action can be taken to overcome or minimize these constraints. The key components of Precision Farming include capturing the data at an appropriate scale and frequency, interpretation and analysis of that data, and implementation of a management response at an appropriate scale and time. In fact, the key differentiator between the conventional management system and precision agriculture is the application of modern information technologies towards providing, processing and analysing the multi-source data of high spatial and temporal resolution for decision-making and operations in the management of crop production.

3.4.2.1 GPS Role in Precision Agriculture

Precision Agriculture is doing the right thing, at the right place, at the right time. Knowing the right thing to do may involve all kinds of high-tech equipment and fancy statistics or other analysis. Doing the right thing however starts with good managers and good operators doing a good job of using common tools such as planters, fertilizer applicators, harvesters and whatever else might be needed. (Colvin and Kerkman,1999).

The use of GPS in Agriculture is limited but it is fair to expect wide spread use of GPS in future. Recently a GPS-based crop duster (precision GPS Helicopter), which can spray an area as small as 4 X 4 meter. is attracting great attention. Some progressive farmers are now beginning to use GPS for recording observations. Such as weed growth, unusual plant stress, colouring and growth conditions, which can then be mapped with a GIS programmes. In the years to come, GPS system role in precision agriculture may help the Indian farmers to harvest the fruits of frontier technologies without compromising the quality of land and produce.

3.4.2.2 Status of Precision Farming in India

The first thing that comes to mind is that, this system is not for developing countries, especially India, where the farmers are poor, farming is mostly subsistent and the land holding size is small. But this is far from the truth as this approach has a large potential for improving the agricultural production in developing world. Imagine this situation where a farmer goes to his field with a GPS (Global Positioning System)-guided tractor. The GPS senses the exact location of tractor within the field. It sends signals to the computer fixed on to the tractor, which has a Geographical Information System (GIS), storing the soil nutrient requirement map in it. The GIS, in consultation with a Decision Support System would decide what the exact requirement of fertilizers for that location is. It then commands a variable rate fertilizer applicator, which is again attached with the tractor, to apply the exact dosage at the precise location of farm. But, this is what precision farming means to large growers in the highly developed parts of the globe. To make it clearer, Precision Farming is the system of matching of resource application and agronomic practices with soil attributes and crop requirements since they vary across a field.

Tata Kisan Kendra: The concept of precision farming being implemented by the TKKs has the potential to catapult rural India from the bullock-cart age into the new era of satellites and IT. TCL's extension services, brought to farmers through the TKKs, use remote-sensing technology to analyse soil, inform about crop health, pest attacks and coverage of various crops predicting the final output. This helps farmers adapt quickly to changing conditions. The result: healthier crops, higher yields and enhanced incomes for farmers.

Government Organization: Precision Agriculture models are not complete, unless the parameters related to empowerment of the farmers; especially small and marginal farmers are integrated. Now it is the turn of good news to the Indian farming community. Some of the research institutes such as Space Applications Centre (ISRO), M.S. Swaminathan Research Foundation, Chennai, Indian Agricultural Research Institute, New Delhi, and Project Directorate of Cropping Systems Research, Modipuram, had started working in this direction and in soon it will help the Indian farmers harvest the fruits of frontier technologies without compromising on the quality of land. According to the Exim Bank officials, though the research and development on PF is currently at a nascent stage in the country, the efforts being put on by the four research institutes were expected to turn the green revolution into an evergreen revolution. In this context, ISRO has also initiated Gramsat project in Orissa. In the line of JDCP, the Gramsat project aims at empowering the people especially the poor and marginalized, by awareness building and access to information and services. Towards this, a network of one-way video and two-way audio Forecasting the yield of mono and multiple crops is being done at NRSA.

Acreage estimates and crop inventory is being done during Kharif and Rabi seasons for Rice, which is the major crop grown in our India. Other crops like Banana, Chillies, Cotton, Maize, Sugarcane and Tobacco are also being inventoried. Satellite data can also delineate different crops that are growing the same area, and an inventory of each of the crops can be done.

3.4.2.3 Precision Farming in India

Precision farming needs information about mean characteristics of small, relatively homogeneous management zones. This is for Data acquisition of the farms to find the soil, vegetation and other parameters that are amenable for remote sensing. Remote sensing techniques play an important role in precision farming by providing continuous acquired data of agricultural crops. Remote sensors image vegetation, which is growing on different soil types with different water availability, substrate, impact of cultivation, and relief Sensors use for the following applications:

- Soil Properties Sensing: Soil Texture, Structure, and Physical Condition; Soil Moisture and Soil Nutrients.
- Crop Sensing: Plant Population; Crop Stress and Nutrient Status.
- Yield Monitoring Systems: Crop Yield; Harvest Swath Width; Crop Moisture:
- Variable Rate Technology Systems: Fertilizer flow; Weed detection, pressure sensors.

Economic feasibility: Unlike some new technologies, there is no clear answer as to whether or not PA is economical beneficial. Indian agriculture conditions.

1. On the one hand there are depletions of ecological foundations of the agro-ecosystems, as reflected in terms of increasing land degradation, depletion of water resources and rising trends of floods, drought and crop pests and diseases. On the other hand, there is imperative socio-economic need to have enhanced productivity per units of land, water and time.
2. At present, 3 ha of rain fed areas produce cereal grain equivalent to that produced in 1 ha. of irrigated. Out of 142 million ha. Net sown areas, 92 million ha. are under rain-fed agriculture in the county.

3. From equity point of view, even the record agricultural production of more than 200 Mt is unable to address food security issue. A close to 60 Mt food grains in the storehouses of Food Corporation of India (FCI) is beyond the affordability and access to the poor and marginalized in many pockets of the country.
4. Globally, there are challenges arising from the Globalization especially the impact of WTO regime on small and marginalized farmers.
5. Some other unforeseen challenges could be anticipated global warming scenario and its possible impact on diverse agro-ecosystems in terms of alterations in traditional crop belts, micro-level perturbations in hydrologic cycle and more uncertain crop-weather interactions etc.

While some studies have reported positive returns to variable-rate technology (VRT), others have reported costs higher than returns or no significant difference in returns.

- Precision Agriculture is a system, not a single piece of equipment or technology. A GPS by itself has little value to farmer. However, when combined with a yield monitor or a VRT, it may have value.
- Returns may be positive if costs can be spread over many applications. Specialized equipment, which has limited uses, has greater risks associated with it than equipment that has many uses. A multi-use tractor will likely pay for itself sooner than a new, single-use machine.
- Precision agriculture may not return on low-valued commodities as it does on high-valued specialty crops i.e., high revenue for grape than the wheat and paddy.

GPS controlled tractor guidance systems may affect when and how tractors are operated. Precision farming is useful in many situations in developing countries. Rice, wheat, sugar beet, onion, potato and cotton among the field crops and apple, grape, tea, coffee and oil palm among horticultural crops are perhaps the most relevant. Some have a very high value per acre, making excellent cases for site-specific management. For all these crops, yield mapping is the first step to determine the precise locations of the highest and lowest yield areas of the field. Researchers at Kyoto University recently developed a two-row rice harvester for determining yields on a micro plot basis (Iida et al., 1998). Subsidies on inputs and outputs and mechanisms that prevent the price system from rationing limited resources are also common. Inputs such as water and fossil fuels are usually sold at prices that are well below the real resource cost of their use, which consists not only production costs but also includes scarcity value and costs of pollution. In such cases, the formulation of policies that reflect the real scarcity value of natural resources and penalize pollution and policies such as green payments for farmers adopting techniques that would lower environmental costs can promote the adoption of precision farming technologies (Branden et al., 1994).

3.5 Chapter Conclusion

This chapter presents basic concepts which pertain to describe agricultural attributes. After literature review, we have found some major technologies and practices are need to define and formularize. The chapter contains definitions, salient features, functions, procedure and methodology of various concepts used to build a data warehouse for Indian agriculture. The chapter helps to understand the problems of Indian agriculture system.

The other useful concepts related to data warehouse design, and DSS are included in the chapter. This chapter also cover the methods and tools used in data mining for large databases. Two data mining techniques: neural network and decision tree are discussed, and used in this study. Chapter also describes the concept of agent technology and multi agent system, and finally in forth section we have given the procedure to visualize data warehouse using geographical information system.

CHAPTER 4

Multidimensional Schema for Agricultural Data Warehouse

In data warehouse literature multidimensional cube and their correspondent star and snowflake schema are used to design subject oriented databases. Dimensional modelling is the most suitable approach to design a data warehouse for the purpose of predictive data mining. According to [Kor99], the objectives of dimensional modelling are: (i) to produce database structures that are easy for end-users to understand and write queries against, and (ii) to maximize the efficiency of queries (Kimball, R., Reeves 1998). One way to look at the multidimensional data model is to view as a cube, the caveat here is that, as the number of dimensional increases, the number of cube's cells increases exponentially (Paulraj Ponniah 2005) On the other hand, the majority of multidimensional queries deal with consolidated and high level of data. Therefore, the solution to building an efficient multidimensional data base is to consolidate all logical attributes. The consolidation is especially valuable since typical dimensional are hierarchical in nature. We found this approach relevant and useful for our work.

The objective of this chapter is to provide a design a multidimensional schema for agriculture data warehouse by placing crops, soil and other parameters in appropriate place. In this chapter the multidimensional schema is explained, that further promotes star schema and snowflake schema. The proposed data warehouse may be extended to Decision Support System (DSS) combine with data mining techniques.

4.1 Data Warehouse Design

The first step in data warehouse design is to organize raw facts into suitable data model. The requirements definition of any data warehouse is completely drives the data design for the data warehouse. Data design consists of putting all the data structure together. The agricultural-data contains a large number of raw facts that are inter- dependent to each other. These facts required processing for converting informational data into operational data. The raw data is first classified in different attributes and then placed in a logical data model. This includes many data warehouse functions such as cleaning, summarizing and reformatting of attributes. An efficient data warehouse must be emphasis on information retrieval techniques. Here efficiency concerns both algorithms and data structured used. Logical data design includes determination of the various data elements that are needed for a particular subject. It also establishes the relationships among various data structures used for storage the data. In this work the multidimensional modelling concepts are used to design agricultural data warehouse. It is a logical design technique to give structure to the agricultural dimensions and metrics, which are analysed as important attributes for agriculture. We attempt to explore eight types of land, depends on combinatory percentage of various fertilizers. We also classify crops in twelve major classes and propose a classifier for weather, based on months of the calendar. These attributes are well placed in the model that systematically evaluates the performance of crops for a particular land, weather and environment domain.

4.2 Data Mining Models

Data mining involves many different algorithms to accomplish different task. The algorithms examine the data and determine a model that is closest to the characteristics of the data being examined· Data mining techniques can be used to make three kinds of model correspondent to three types of task: descriptive model, profiling model, and predictive model. Hypothesis testing often produces descriptive models. On the other hand, both profiling model and predictive model have a goal in mind when a model is being built.

4.2.1 Descriptive Models

Descriptive model describes what is in the data. The output in one or more charts or numbers or graphics that explain what is going on. A descriptive model identifies patterns or relationships in data; it serves as a way to explore the properties of data examined; not to predict the new properties. Clustering, summarization, association rules, and sequence discovery are usually viewed as descriptive in nature. Clustering, summarization, association rules, and sequence discovery are usually viewed as descriptive in nature. Clustering, summarization, association rules, and sequence discovery are usually viewed as descriptive in nature.

4.2.2 Profiling Models

Profiling models are often based on demographic variables. In profiling models, the target is from the same time frame as the input. Profiling model has serious limitations. One is the inability to distinguish cause and effect, as the profiling is based on familiar demographic variables. It needs not to involve any sophisticated data analysis. Surveys are the common method of building the profiles. The distinction between profiling and prediction is that profiling has implications for the modelling methodology, especially the treatment of time in the creation of the model set.

4.2.3 Predictive Models

A predictive model makes a prediction about values of data using known results found from different data sources. Prediction means finding pattern in data from one period that are capable of explaining outcome in a later period. Predictive modelling may be made based on the use of other historical data, and it predicts what is likely to happen in the future. Building a predictive model requires separation in time between the model inputs and the model output, the thing to be predicted. Predictive model data mining tasks includes classification, regression, time series analysis, and prediction. In our study we use predictive approach to predict performance of crops in a particular time and particular type of soil and fertilizer set.

4.3 Multidimensional Schema for Agricultural Data Warehouse

Dimensional modelling is a different way to view and interrogate data in database. This view may be used in a DSS in conjunction with data mining tasks. Decision support applications often require that information obtained along many dimensions. For example, a farmer may want to predict production of any crop, in a specific geographic region, in particular month for seeding, and by crop type. This query requires three dimensions: Crop-set, Seeding Month, Land type. Each dimension is a collection of logically related attributes and is viewed as an axis for modelling the data. Within each dimension, these entities form levels, on which various questions may be asked. The specific data stored is called facts and usually numeric data.

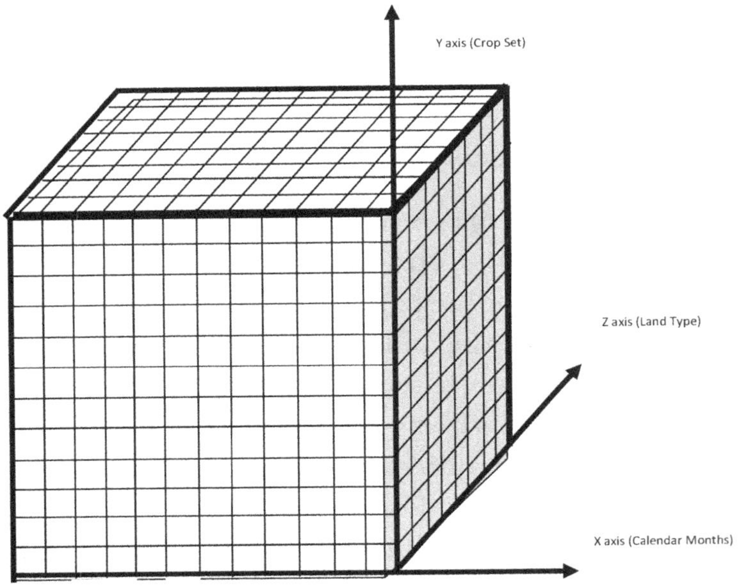

Figure: 4.3- Multidimensional Cube for Agricultural Attributes

In our schema "12 months" (Jan to Dec) are the main classifier of weather and placed along x- axis. The attribute "crop-sets" are placed along y-axis and "land-type", we have placed along z-axis. Let m is the number of calendar months placed along x-axis, c is the number of crop-sets at y-axis and l is the number of land type placed at z -axis. Then total number of values in cube is given by $V_n = m \times c \times l$ i.e., $12 \times 12 \times 8 = 1152$. The detail description of cube is given in following sections.

4.3.1 Calendar for Seeding of Crops

Every crop needs particular and specific weather for the best results. In our multidimensional schema 12-month calendar is used as a main classifier of weather. We have assumed that maximum production of any crop is possible when it is seeded in suitable month of the calendar. The weather report for every month of calendar is based on some factors, such as cold, rain, humidity and heat. Forecasting about weather is dependent to all these factors, table 4.4.1 shows predicted percentage of cold, rain, humidity and rain in large part on India in corresponding to months of the calendar.

Table: 4.3.1 Weather parameters correspondent to calendar month in major part of India

	Month	% Cold	% Rain	% Humidity	% Heat
1	January	60	15	20	5
2	February	60	15	15	10
3	March	50	5	10	35
4	April	25	5	10	60
5	May	15	10	5	70
6	June	5	30	5	60
7	July	5	55	25	15
8	August	10	60	25	5
9	September	10	60	25	5
10	October	15	50	25	10
11	November	35	25	30	10
12	December	50	15	30	5

4.3.2 Crop-Sets

There are hundreds of crops whose success level depends on some attributes such as start time, ending time, requirements of fertilizers and type of seeds. We identifying 12 major crop sets correspond to their seeding months. These crop set consists number of crops that require similar weather and environmental needs. As displays in table 4.4.2, similar types of crops are put together in a crop-set and identified by the name of major crop belongs to it.

- Wheat

- Potatoes
- Mustered
- Oil Seed
- Coffee
- Pulses
- Rice
- Peppermint
- Sunflower
- Cash crops
- Coconut
- Sugarcane

Table: 4.3.2- Major crops of India by their starting months

Crop Set	Crop	Starting Month	Ending Month
I	Wheat	December	March
II	Potato	November	January
III	Mustered	October	March
IV	Oil-seed	September	February
V	Coffee	August	November
VI	Pulses	July	October
VII	Rice	June	October
VIII	Peppermint	May	July
IX	Sunflower	April	June
X	Cash-crops	March	June
XI	Coconut	February	August
XII	Sugarcane	January	September

4.3.3 Land Types

The impact of fertilizers on production of any crop is significantly measured. In a particular type of land, there are various fertilizers and minerals. We can identify many types of land based on combinatory ratio of various fertilizers in soil. In our study we identified 8 major fertilizers that are naturally available in soil. However, we can generalize the fertilizer-set, which may change with geographical regions.

1. Urea (U)
2. Phosphorus (P)
3. Iron (Fe)
4. Nitrogen (N)
5. Zinc(Z)
6. Sulfur (S)
7. Calcium (Ca)
8. Potassium (K)

4.3.3.1 Fertilizer-sets

In a particular type of soil, there are various fertilizers and minerals. The impact of fertilizers on production of any crop is significantly measured. We can identify various types of soil based on combinatory ratio of these fertilizers. In our study we identified 8 major fertilizers that are naturally available in soil, for this study fertilizer-set FS is defined as: FS = {Urea (U), Phosphorus (P), Iron (Fe), Nitrogen (N), Zinc (Z), Sulfur (S), Calcium (Ca), Potassium (K)}. Further on the basis of the ratio of above fertilizers presence in soil, we can identify other soil types. Following table displays classification of soil type depends on percentage of fertilizers in a soil. In this table we hypothetically demonstrate 8 different type of soil based on fertilizer-set FS, however for the generalization purpose we can identify n types of soil, depending on combinatory ratio of fertilizers-set FS. We can also identify other fertilizer-sets, that may has other elements as defined in the set FS i.e., for any fertilizer-set where FS ≠ {Urea (U), Phosphorus (P), Iron (Fe), Nitrogen (N), Zinc(Z), Sulfur (S), Calcium (Ca), Potassium (K)}. By this way we can generalize the fertilizer-set, which may change with geographical regions.

Table: 4.3.3.1 Land Type basis on fertilizers set in soil

S No.	Land Type	Percentage of Fertilizers in Fertilizer Set								Weight of Fertilizer	Total Weight
		U	P	Fe	N	Z	S	Ca	K	(P_f)	(F_w)
1	L1	20	-	-	50	-	-	-	10	U2 + N5 + K1	8
2	L2	20	30	10	10	10	-	-	-	U2 + P3 + Fe1 + N1 + Z1	8
3	L3	20		10	20	10	20	-	-	U2 + Fe1 + N2 + Z1 + S2	8
4	L4	-	30	-	20	-	-	10	20	P3 + N2 + Ca1 + K2	8
5	L5	10	-	-	30	-	10	20	10	U1 + N3 + S1+Ca2 + K1	8
6	L6	40	10	-	10	-	20	-	-	U4 + P1 + N1 + S2	8
7	L7	-	-	40	10	20	-	10	-	I4 + N1 + Z2 + Ca1	8
8	L8	10	10	10	20	-	30	-	-	U1 + P1 + Fe1 + N2 + S3	8

Further on the basis of the ratio of above fertilizers present in soil, we have identified 8 types of land for our study; table 4.4.3 displays land type depends on percentage of fertilizers available in a soil. In table 4.4.3 we hypothetically demonstrate 8 different type of land based on above fertilizer, however for the generalization purpose we can identify n types of land. In the model, further we assign 8-unit weight to each type of soil; each unit of weight corresponds to 10 percent availability of a particular fertilizer in any fertilizer-set. For example, land type L1 contains 50 percent of Nitrogen, 20 percent of Urea and 10 percent of Potassium (table 4.4.3) and, rest 20 percent includes other elements in soil. We are not including such 20 percent in the study, assuming it is non-deterministic minerals, as soil has many other impurities.

4.3.4 Irrigation

Irrigation is an important factor that majorly affects production of any crop. In our model irrigation factor is consider as a dimensional of the cube (not shown in figure). We have taken three possible states of irrigation: high irrigation, medium irrigation and low irrigation. Accordingly, we assign some constant values to each of these levels of irrigation, for the computational purpose. The irrigation factor is also described in correspondent star schema and further extended into snowflake schema.

4.3.5 Seed Quality

With the development of biotechnology, researcher produces high quality of seeds; the production rate of any crop is certainly dependent on quality of seed. As we have done for irrigation, quality of seeds is also classified in three levels, high quality, medium quality and low quality. Further we have assigned some constant values to each of these levels of seed quality in our model. Like irrigation seed quality is also consider as a dimension of the cube (not shown in figure). The Seed quality factor is also described in correspondent star schema and then, in snowflake schema.

4.4 Arranging Dimensions in Schema

Specialized schemas have been developed to portray multidimensional data. These include star schema and snowflake schema. A star schema shows data as a collection of two types: facts and dimensions. The star schema view can be obtained via a relation system where each dimension is a table and the facts are stored in fact table. Unlike relation schema which is flat, a star schema is a graphical view of data. At the center of star, the data being examined, the facts are shown in the fact table. On the outside of the facts, each dimension is shown separately in dimension table.

Descriptive information about the dimensions is stored in dimensions tables, which tend to be smaller. While the actual data is being accessed, are stored in fact table and thus tend to be quite large. To ensure efficiency of access, facts may be stored for all possible aggregation levels. The fact data would then be extended to include a level indicator. Snowflake schema is an extension of star schema; it facilitates more complex data views. In snowflake schema, the aggregation hierarchy is described explicitly in the schema itself. The snowflake schema can be understood as partially normalized version of corresponding star schema.

4.4.1 Star Schema Arrangement

The star schema proposed for multidimensional agricultural data warehouse is starts with a central fact table that corresponds to facts about agricultural technology. Data in fact table can be viewed as a regular relation with an attribute for each fact, to be stored and the key being the values, for each dimension. Each row of the central table contains some combination of keys that makes a fact unique. These keys are called dimensions.

In our model we use simplest star schema that has one fact table and three-dimensional tables corresponds to weather, crop-sets and land type, two other dimensions, irrigation and seed quality is also included in the model for computation of decision coefficient. The central fact table also has other columns that typically contain information specific to each row, such as irrigation level, sequence of crop-set and seeding month and, quality of seed. For the purpose of computation, we have assigned numeric values to all this information. Figure 4.4.1 explains the star schema arrangements of multidimensional cube given in previous section. The fact table contains major attributes such as irrigation level, crop-set, seeding month of the crop, soil type and quality of seed.

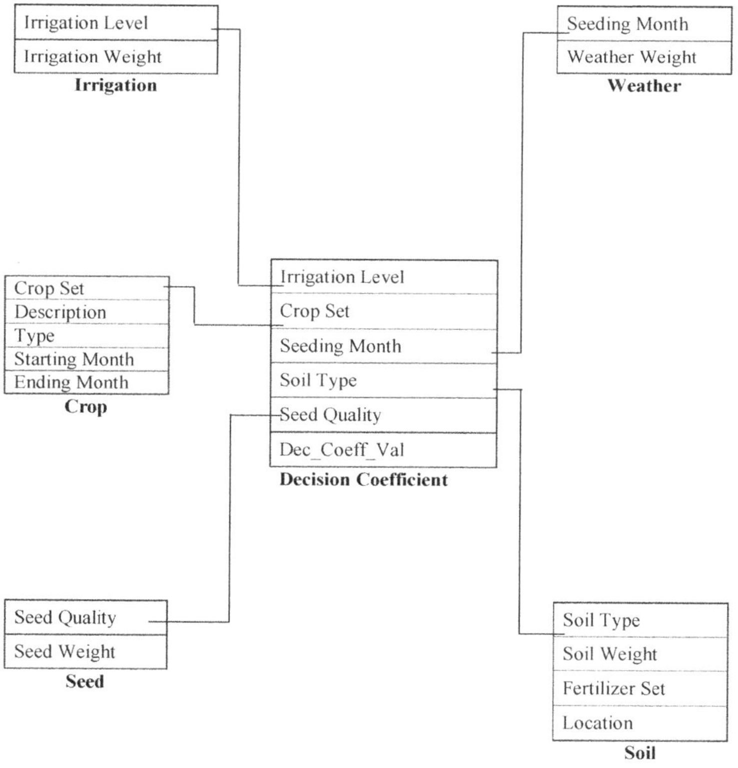

Figure: 4.4.1- Star Schema for Agricultural Data Warehouse

A computational model is required to produce the value of decision coefficient. Decision coefficient is the success ratio of any crop based on above attributes describe in the cube. We can compute the value of decision coefficient for any crop by providing values of these attributes as input, to the computational model. There are four basic approaches to the storage of data in dimensional table [PB99], which are flattened approach, normalized approach expanded approach and levelized approach. Each dimension table can be stored in one of these manners.

This study has adapted normalized approach to store the data and to remove duplicate values. In this approach a table exists for each level in each dimension. Each table has one tuple for every occurrence at that level. As with normalization each lower-level dimension table has a foreign key pointing to next higher level. Figure 4.4.2 provides queries to implement star schema for agricultural data warehouse using normalized approach.

4.4.2 Snowflake Schema Arrangement

A star schema stores all attributes for a dimension into one de-normalized (flattened) table. This requires more disk space than a snowflake schema that is eventually more normalized approach. Snowflake schema normalizes the dimension by moving attributes with few distinct values into separate dimension tables that relates to core dimension table by using foreign keys.

A snowflake schema is a logical arrangement of tables in a multidimensional database, and so the entity relationship diagram resembles as a snowflake in shape. The snowflake schema is represented by centralized fact tables which are connected to multiple dimensions. In snowflake schema, dimensions are normalized into multiple related tables, where in star schema dimensions are normalized with each dimension represented by a single table.

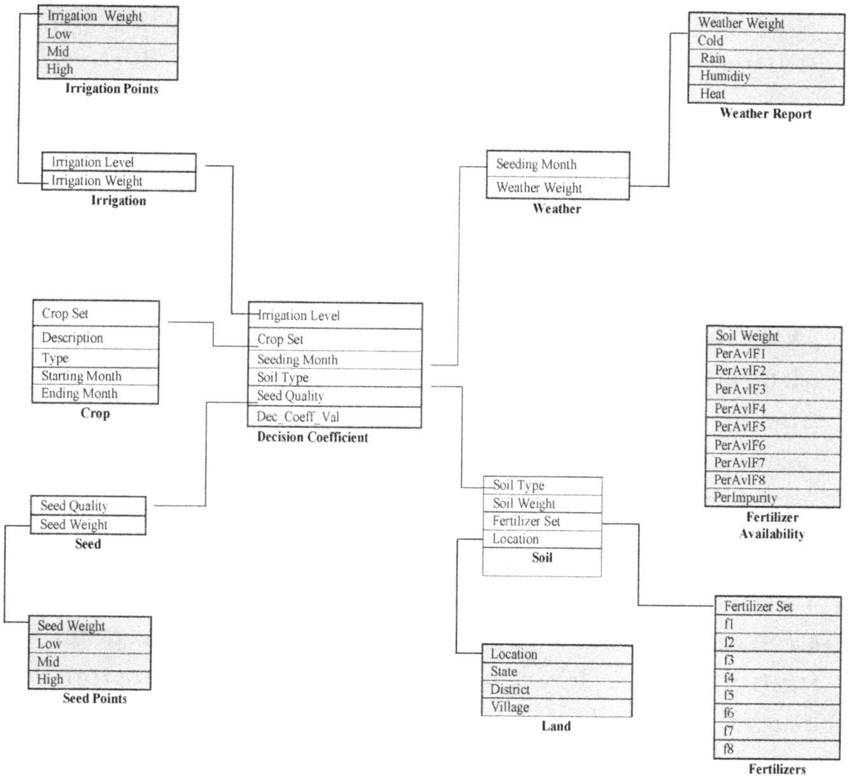

Figure: 4.4.2- Extension of Star Schema into Snowflake Schema for Agricultural Data Warehouse

Each dimension table is in a one-to-many relationship with the central fact table. So, the primary key of each dimension table must be a foreign key in the fact table. If there are five dimensions crop, seed, weather, irrigation and soil type, then the primary key of each of these tables must be presented in the orders fact table as foreign keys. There are three options.

1. **A single compound primary key whose length is total length of the keys of the individual dimensional tables.** Under this option, in addition to compound primary key,

the foreign keys must also be kept in fact table as additional attributes. This option increases the size of table.

2. **Concatenated primary key that is concatenation of all primary keys of the dimension tables.** Here we need not to keep the primary keys of dimension tables as additional attributes to serve as foreign keys. The individual parts of primary keys themselves will serve as the foreign keys.

3. **A general primary key independent of the keys of dimension tables.** In addition to generated primary key, the foreign keys must also be kept in the fact table as additional attributes. This option also increases size of the fact table.

A complex snowflake shape emerges when the dimensions of a star schema are elaborated, having multiple levels of relationships, i.e. the child tables have multiple parent tables. In figure 4.4.2 we have elaborated weather dimension into weather report, irrigation into irrigation points, and seed into seed points.

Decision Coefficient (**Irrigation Level, Crop Set, Seeding Month, Soil Type, Seed Quality,** *Dec_Coeff_Val)*

Irrigation (**Irrigation Level,** *Irrigation Weight)*

Irrigation Points (**Irrigation Weight**, *Low, Mid, High)*

Crop (**Crop Set**, *Description, Type, Starting Month, Ending Month)*

Weather (**Seeding Month, Weather Weight**)

Weather Report (**Weather Weight**, *Cold, Rain, Humidity, Heat)*

Soil (**Soil Type, Soil Weight, Fertilizer Set**, *Location)*

Fertilizers Availability (**Soil weight,** *PerAvlF1, PerAvlF2, PerAvlF3, PerAvlF4, PerAvlF5, PerAvlF6, PerAvlF7, PerAvlF8, PerImpurity)*

Fertilizer (**Fertilizer Set**, *f1, f2, f2, f4, f5, f6, f7, f8)*

Land (**Location**, *State, District, Village)*

Seed (**Seed Quality**, *Seed Weight)*

Seed Points (**Seed Weight**, *Low, Mid, High)*

Figure: 4.5.2.1- Schema definition of correspond to snowflake schema

The soil dimensions are extended into three relationships i.e., land, fertilizer availability and fertilizer weight. Snowflake is used to improve query performance against low cardinality attributes. Business intelligence applications that use OLAP architecture may perform better, when the snowflake schema is used for the purpose of data warehousing.

4.5 OLAP for Agricultural Data Warehouse

Online analytical processing system are targeted to provide more complex query results than online transaction processing system (OLTP) or relational database system. The OLAP system provides extra analysis of data as well as more imprecise nature of queries. This basically differentiates an OLAP application from traditional database or OLTP application. The multidimensional view of data is fundamental to OLAP applications. The complex nature of OLAP application requires a multidimensional view of data, and the type of data accessed is often a data warehouse. OLAP tools can be classified as relational OLAP or multidimensional OLAP (MOLAP). The data warehouse systems often contain multiple OLAP cubes, and the power of OLAP arises from practice of sharing dimensional across different cubes.

In MOLAP system, data are modelled, viewed, and physically stored in a multidimensional schema. MOLAP tools are implemented by specialized DBMS and software systems capable of supporting the multidimensional data. With MOLAP the cube view of data is stored as an n- dimensional array, this approach requires extremely high storage space and then indices may be used to speed up processing of data. The parallel computing can be used to overcome this limitation, and then such cube requires less processing time.

4.5.1 Drill Down and Rolling Up

In our agricultural data warehouse, we purposefully use multi-dimensional cube to store the data. Here every cube is particularly valuable because they make possible to drill-down and roll-up operations with other cubes. To assist roll-up and drill down operations aggregation is used, and the values that are pre-computed are stored in the model. The attributes in a dimension table provide the ability to get the details from higher levels of aggregation to lower levels of details. For three attributes Location state, district, village form a hierarchy. We can get the details of soil attributes by state wise, district wise and village wise. We may first get village then district and then state to roll up the information. In section 4.5.1 it is mentioned the requirement of a computational model, which computes value of decision coefficient given in star schema.

A simple query may look at a single cell within the cube; however, queries may state in multidimensional terms. There are several types of OLAP operations that also support to response more complex queries.

Slice: This operation performed by selecting values on one dimension. Such as, values of decision coefficient for all crops in a particular month. The SQL statement may be given as.

SELECT ALL (Dec_Coeff_Val)
FROM Decision_Coefficient
WHERE Seeding_Month = 'July';

Dice: This can be performed by slice on one dimension and then rotating the cube to select on second dimension i.e.

SELECT Dec_Coeff_Val
FROM Decision_Coefficient
WHERE

Seeding_Month **IN** ('Jan', 'Feb', 'March', 'April', 'May', 'June', 'July', 'Aug', 'Sep', 'Oct', 'Nov' , 'Dec');

Roll-up: This allows asking queries that moves up an aggregation hierarchy. Instead of looking one fact we look at all the facts.

SELECT ALL (Dec_Coeff_Val)
FROM Decision_Coefficient
WHERE

Seeding_Month **IN** ('Jan', 'Feb', 'March', 'April', 'May', 'June', 'July', 'Aug', 'Sep', 'Oct', 'Nov', 'Dec');

Drill-down: This operation allows user to navigate lower in the aggregation hierarchy. In this user get more specified results. Such as, value of decision coefficient for a particular crop in a particular month.

SELECT Dec_Coeff_Val
FROM Decision_Coefficient
WHERE Seeding_Month = 'July';

The other OLAP operations may be used to support more analytical results of agricultural facts. We can combine each of above OLAP operations to response more analytical queries.

4.6 Chapter Conclusion

In this chapter we have applied concepts of dimensional modelling. We have identified major agricultural attributes and place these attributes in the correspondent dimensions of the cube. Further we developed a star schema for multidimensional data cube and identified the primary keys to connect each of the dimensions with central fact table of the star schema. The fact table collects information from each of the dimensions and then, the values of decision coefficient is computed and stored in the fact table, correspondent to each cell of the cube. We have also extended the star schema in to snowflake schema by applying hierarchical aggregation of each dimension. In snowflake schema, the microscopic view of factors, affecting each dimension of the cube is clearly explained. In the last section of the work, we have applied OLAP operations to analyze the facts stored in the multidimensional database. The OLAP tools help farmers and decision makers to predict production of crops, to find the lack of any fertilizer available in soil or to guess the level of irrigation required, during growth of the crop.

The real strength of this chapter lies in the adoption of predictive data mining techniques. Many data mining tools and algorithm may be applied to this agro database, which probably responds to more complex queries. For example, a farmer might discover that a particular crop produces better results, when it is irrigated at a particular time, under a particular weather condition. The decision maker may find some pattern in crop sequence for better results such as farmers of Chhattisgarh, Orissa and parts of Bihar could found that rice-wheat system is more suitable than rice-pulses system. Similarly, other complex and specific queries can respond using data mining. This manuscript strongly advocates the use of predictive data mining techniques to retrieve more sophisticated information from agricultural data warehouse.

In the next chapter we will provide a computational model that is used to compute the values in multidimensional cube. Further in other chapter we provide concept and formulization for agent-based software based on data mining techniques.

CHAPTER 5

Computational Model for Agricultural Decision Support System

An expert system for use in land drainage decisions was designed to diagnose the causes of drainage problems in the command area of an irrigation system (N. Haie and R.W. Irwin. 1988). Factors such as water regime in the soil profile, presence of a cultivation pan or an impermeable layer below the topsoil etc., were considered. An expert system for crop variety selection was developed for winter wheat in Scotland (O. W. Morgan, M.J. McGregor, M. Richards and K.E. Oskouri. 1989). The developed system was designed to consider soil characteristics, water availability and prevalence of diseases.

In the previous chapter we introduced a multidimensional schema for agriculture data ware house. That leads to many interdependencies between dimension tables and fact table, explained in snowflake schema. In the absence of computational model, an enormous amount of redundancy of data is observed, it increases complexities in OLAP. To overcome this redundancy in operational data, a mathematical procedure is needed for computation of decision coefficient. In this chapter, we are introducing a computational model to optimize cropping system that leads to design of data warehouse for agriculture. This chapter also led to design an agriculture data warehouse and provide a decision support system, based on OLAP and data mining techniques.

5.1 Gathering Agricultural Data

Agricultural data is diversified and complex. It has large volume and many inter-related attributes. Extracting useful information and, decision making on this large volume of data is difficult and costly. As a result, we require an efficient model that stores agricultural data in proper manner and efficient to response ad-hoc queries of the farmers and economist. In particular, agriculture technology is dependent on the prediction about weather and diagnosis of fertilizers in soil, type of crop and environmental issues. Too precisely for particular crop, we have to select proper quantity of various minerals and fertilizers related to a particular soil. Similarly, the other attributes related to environment and weather may be classified on the basis of cold, rain, humidity, heat and other issues. The optimal selections of all these attributes help us to predict the production of crop and diagnose all related agricultural issues. The proposed model produces mathematical values of decision-making coefficient for production of various types of crop, in a particular land type and in a particular environmental attribute. The mathematical equations and supported tables are included to rectify the purpose of the model.

5.2 Computational Model for Agricultural Decision Support System

In privies chapter we have explored eight types of soil, depends on combinatory percentage of various fertilizers. We also classified crops in twelve major classes and propose a classifier for weather based on calendar months. These attributes are well placed in the multidimensional cube. To evaluate performance of crops for a particular soil, weather and for other agricultural attributes we required a computational model, which calculate the decision coefficient and store it in multidimensional schema. The major inputs of the model are described in the following sections.

5.2.1 Calendar Months

Every crop needs specific weather for the best result; therefore, the success ratio of a particular crop is depended on its seeding time. In our mathematical model, we use 12 months calendar as a main classifier of weather (also explain in previous chapter), we hypothetically assume maximum yield of a crop is possible when it is sowing in suitable month of the calendar. Although, forecasting about weather is unpredicted and dependent to some factors such as cold, rain, humidity and heat. (See details in previous chapter table 4.3.1)

5.2.2 Crop-Sets

There are hundreds of crops, and their success level depends on some attributes such as start time, ending time, requirements of fertilizers and type of seeds. In our model similar types of crops are put together. We identify twelve major crop-sets corresponding to 12 months of calendar by their seeding time. These crop-sets consists number of crops having similar weather and environmental needs. These crops are highly related to a particular class, and every crop belongs to particular crop-set. Here, we are putting similar crops in a set. These crop sets may be identified by the name of major crop belongs to it.

5.2.3 Fertilizer-Sets

In a particular type of soil, there are various fertilizers and minerals. The impact of fertilizers on production of any crop is significantly measured.

Table: 5.2.3- Land Type basis on fertilizers set in soil (Repeated)

S No.	Land Type	Percentage of Fertilizers in Fertilizer Set								Weight of Fertilizer	Total Weight
		U	P	Fe	N	Z	S	Ca	K	(P_f)	(F_w)
1	L1	20	-	-	50	-	-	-	10	U2 + N5 + K1	8
2	L2	20	30	10	10	10	-	-	-	U2 + P3 + Fe1+N1+ Z1	8
3	L3	20		10	20	10	20	-	-	U2 + Fe1 + N2 + Z1+ S2	8
4	L4	-	30	-	20	-	-	10	20	P3 + N2 + Ca1+ K2	8
5	L5	10	-	-	30	-	10	20	10	U1 + N3 + S1+Ca2 + K1	8
6	L6	40	10	-	10	-	20	-	-	U4 + P1 + N1 + S2	8
7	L7	-	-	40	10	20	-	10	-	I4 + N1 + Z2 + Ca1	8
8	L8	10	10	10	20	-	30	-	-	U1 + P1 + Fe1 + N2 + S3	8

We can identify various types of soil based on combinatory ratio of these fertilizers. In our study we identified 8 major fertilizers that are naturally available in soil, for this study fertilizer-set FS is defined as: FS = {Urea (U), Phosphorus (P), Iron (Fe), Nitrogen (N), Zinc (Z), Sulfur (S), Calcium (Ca), Potassium (K)}. Further on the basis of the ratio of above fertilizers presence in soil, we can identify other soil types. For the convenience we are giving the table explained in section 4.4.3

In the model, further we assign 8-unit weight to each type of soil; each unit of weight corresponds to 10 % availability of a particular fertilizer in any fertilizer- set. For example, land type L1 contains 20 % of Urea, 50 % of Nitrogen and 10% of Potassium (Table 4.3.3) and, rest 20 % includes other elements in soil. We are not included such 20 % in our study, assuming it is non-deterministic minerals, as soil has many other impurities in it. However, we can identify other types of soil by combining many other fertilizers. Although we have mentioned it in previous chapter then even, we find relevant repeat it here.

Other fertilizer-sets can be identified, that may have other elements as defined in the set FS i.e., for any fertilizer-set where FS ≠ {Urea (U), Phosphorus (P), Iron (Fe), Nitrogen (N), Zinc(Z), Sulfur (S), Calcium (Ca), Potassium (K)}. By this way we can generalize the fertilizer-set, which may change with geographical regions.

5.2.4 Irrigation

In previous chapter we have mention that "Irrigation is an important factor that is also incorporated in our model"; we have taken three possible states of irrigation; high irrigation, medium irrigation and low irrigation. We have assigned constant values to each of these levels of irrigation, for the computational purpose. In our model we assign total 3 units of weight for irrigation factor I_w, the value for irrigation factor $I_w = 3$ if the irrigation is high , irrigation factor $I_w = 2$ if irrigation is medium and , irrigation factor $I_w = 1$ if irrigation is low.

5.2.5 Seed Quality

In previous chapter we have mention "the production rate of any yield is certainly dependent on quality of seed". As we have done for irrigation, quality of seeds is also classified in three levels, high quality, medium quality and low quality. Similarly, the weight of seed-quality S_w is also assign by 3 unit of total weights, and the values of S_w = ,3 for high quality seeds, $S_w = 2$, for medium quality and $S_w = 1$, for low quality of seeds.

5.3 Detail Description of the Model

In this model we introduce a crop cycle based on their starting time and required soil type. The five main attributes: (1) calendar months (2) crop-sets (3) soil-type (4) irrigation and (5) quality of seeds is placed in from inner cycles to outer cycles respectively (figure 5.3.1). Further the model is horizontally dived into in twelve regions each corresponds to month of calendar. The detail description of the model is given as follows:

- The 12 calendar months are main classifier of weather and placed at innermost cycle (cycle 1) in clockwise direction. Every calendar month is represented by its order in calendar year. For example, January is numbered as 1, February as 2 and so on.
- The crop set are identified by major crop belongs to that crop set; these crop set are represented by roman numbers in the model. The crops are placed at cycle 2 in anti-clockwise direction. Each crop set is placed corresponding to their starting month. In our study we have assumed standard starting month of crop-set I is December, and it is placed in month of December. In the same fashion crop-set XII is placed in month of January, crop-set XI is placed in February and so on i.e., the sum of month number and crop set is always 13.
- Cycle 3 corresponds to a particular fertilizer set, and type of soil belongs to that fertilizer set. In our study we identified 8 major fertilizers that are naturally available in soil, further we assign 1 unit weight for every 10 % of availability of a particular fertilizer in a fertilizer set.

By this way, total weight of each soil type is 8 units per section. It means we are measuring only 80% of soil in our study, rest 20% are ignored assuming as non-deterministic data.

- The cycle 4 corresponds to irrigation factor. We have assigned total 3 units weight to irrigation for high value of irrigation, 2 units weights for medium irrigation and for low irrigation weight is equal to 1 unit.
- At outermost cycle, i.e., cycle 5, we have assigned total 3 units weight for high quality of seed, 2 units weight for medium quality of seed and for low quality of seed weight is equal to 1 unit.

5.4 Mathematical Procedure and Working

To demonstrate working of the model, first we consider ideal state of the model; by the ideal state we mean the state in which a particular crop gives best possible production. In other words, if a crop receives most favorable fertilizer- set, most favorable weather with best quality of seed and, also receives proper irrigation, it results best possible production of yield and it is said to be in ideal state. In figure 5.4.1 the model is in ideal state i.e., the crop-sets are inseminated in the proper calendar month and also receives most suitable fertilizers, irrigation and seed quality is high.

Mathematically, if C_j is sequence of crop- set and M_k is the sequence of sowing month for any crop-set, then weight of seeding month and crop-set is given as:

$$M_w = (M + C) \qquad (5.4.1)$$

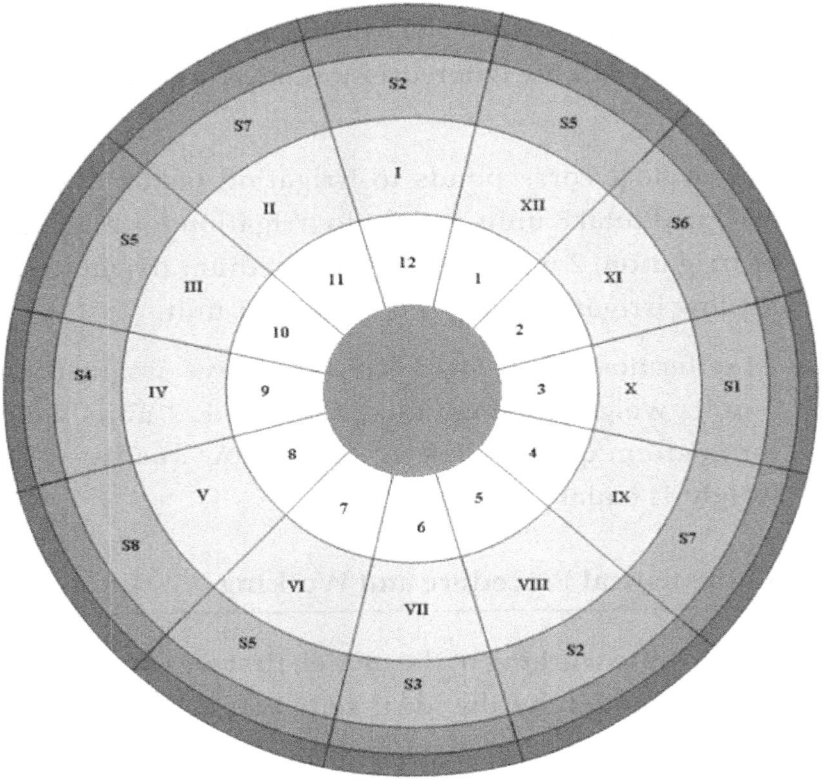

Figure: 5.4 - Computational Model for Agricultural Decision Support System

Explanation Figure 5.4.1				S1 = Soil Type 1
				S2 = Soil Type 2
I = Crop-Set 1	VII = Crop-Set 7	1= January	7= July	S3 = Soil Type 3
II = Crop-Set 2	VIII = Crop-Set 8	2 = February	8 = August	S4 = Soil Type 4
III = Crop-Set 3	IX = Crop-Set 9	3 = March	9 = September	S5 = Soil Type 5
IV = Crop-Set 4	X = Crop-Set 10	4 = April	10 = October	S6 = Soil Type 6
V = Crop-Set 5	XI = Crop-Set 11	5 = May	11 = November	S7 = Soil Type 7
VI = Crop-Set 6	XII = Crop-Set 12	6 = June	12 = December	S8 = Soil Type 8

Where:

M = Sequence of Seeding month in calendar

C = Sequence of crop-set

Let, F_w is the total weight of fertilization factor and I_w, and S_w is the weight of irrigation factors and seed-quality factor respectively, then total weight T_w of any particular crop set for best result is given as:

$$T_w = M_w + F_w + I_w + S_w \qquad (5.4.2)$$

Where:

M_w = Weight of seeding month for any crop set i.e. $M_w = (M_j + C_k)$

F_w = Weight of fertilization factor

I_w = Weight of irrigation factor

S_w = Weight of seed-quality factor

5.4.1 Assigning the Weight

- Observe the values of calendar month cycle and values of crop-set cycle, as the calendar month is kept in clock-wise direction and crop-set are taken place in anti-clock wise direction that resultant, the sum of both of these cycle values is always 13, i.e., $M_w = (M_j + C_k) = 13$.

- Further the total weight of fertilization factor in ideal state is calculated by assuming that all the required fertilizers of any fertilizer–set are available in proper ratio for a particular soil-type. We have already assigned 1-unit weight for every 10% of availability, of a particular fertilizer in the fertilizer set. Hence, the total weight of each soil-type is 8 units per section. Therefore, total weight $F_w = 8$, for ideal state of the model.

- In our model we assign total 3 units of weight for irrigation factor I_w, the value for irrigation factor $I_{w=}$ 3 if the irrigation is high, irrigation factor $I_w = 2$ if irrigation is medium and, irrigation factor $I_{w=}$ 1 if irrigation is low.

- Similarly, the weight of seed-quality S_w is also assign by 3 unit of total weight and the values of $S_w = 3$ for high quality seed, $S_w = 2$ for medium quality and $S_{w=}$ *1* for low quality seeds.

From above considerations, the total weight T_w for ideal state is given by equation (5.4.2)

$T_w = M_w + F_w + I_w + S_w$

$T_w = 13 + 8 + 3 + 3 = 27$ ....(5.4.3)

Computation

Hypothetically, we assume the production of crop is 100% if model score all 27 units of total weight T_w

Let Ω is a multiplicative constant to calculate the value of each point in percent then Ω is be given as:

$\Omega = 100 / 27 = 3.7037$.. (5.4.4)

Therefore, every unit of weight corresponds to 3.7037 percent of success of the crop.

For a particular case, Let P_m is the total points of months, that are obtained by adding the values of cycle 1 and cycle 2, i.e., $P_m = (M_j + C_k)$, where M_j is the sequence of month and C_k is the sequence of crop-set in a particular region.

Similarly, we assume P_f is total points of fertilizers earn in cycle 3, P_i is total points of irrigation factor earn in cycle 4, and finally P_s is the total points of seed-quality earned in the outermost cycle. Then, P_m, P_f, P_i and P_s is calculated as follows:

5.4.2 Points of Month (P_m)

According to the model the sum of month-sequence and crop-set are measured together, for the ideal state total weight of months $M_w = 13$, and if d is the displacement from seeding month of any crop-set from the seeding month in ideal state of the model, then we have

$P_m = M_w \times \Omega - (d \times \Omega)$

$P_m = 13 \times 3.7037 - (d \times \Omega)$ or

$P_m = 48.1481 - (d \times \Omega)$...... .... (5.4.5)

In above equation displacement d is the number of moths between sowing month of any crop, and sowing month in ideal state. Observe the model, any crop belongs to crop-set 1 is should be started in the month of December for the ideal state, suppose it is inseminated in the month of January or November, then the displacement d is equal to 1, and if it is inseminated in the month of October or February the displacement is equal to 2, and so on. Reader may observe this by rotating crop-set cycle on the inner month- cycle, based on this logic displacement can be calculate by following algorithm.

5.4.2.1 Displacement Algorithm

Displacement (M_j, C_k)

// M_j is the sequence of calendar month.

// C_k is the sequence of crop-set

// P_m is the total point of month i.e., $P_m = M_j + C_k$,

// where //$1 \leq j \leq 12$ and $1 \leq k \leq 12$.

// disp is a temporary variable for displacement of any

//crop.

//set from ideal state, d is the output

//P_m, d and, disp are declared as integer variables and

// M_w = 13 initialize.

{

$P_m = M_j + C_k$;
if (P_m = = M_w) disp = 0 ;
if ($P_m < M_w$) disp = P_m –1;
if ($P_m > M_w$) disp = $P_m - T_w$;
if (disp ≤ 6){
    d = disp;
    }
    else {
        d = 12 – disp;
    }
}

With the above algorithm the value of d is between 0 and 6 i.e., $0 \leq d \leq 6$. We now put *d* = (0,1,2,3,4,5,6) in equation (5.4.5), then values of P_m are given as follows:

Based on the above algorithm, following table demonstrate each possible values of *d* for all crop-set given in model.

5.4.2.2 Displacement Table

Table: 5.4.2.2- Values of displacement of particular crop-set in all calendar month

Crop-Set →	I	II	III	IV	V	VI	VII	VIII	IX	X	XI	XII
December	0	1	2	3	4	5	6	5	4	3	2	1
November	1	0	1	2	3	4	5	6	5	4	3	2
October	2	1	0	1	2	3	4	5	6	5	4	3
September	3	2	1	0	1	2	3	4	5	6	5	4
August	4	3	2	1	0	1	2	3	4	5	6	5
July	5	4	3	2	1	0	1	2	3	4	5	6
June	6	5	4	3	2	1	0	1	2	3	4	5
May	5	6	5	4	3	2	1	0	1	2	3	4
April	4	5	6	5	4	3	2	1	0	1	2	3
March	3	4	5	6	5	4	3	2	1	0	1	2
February	2	3	4	5	6	5	4	3	2	1	0	1
January	1	2	3	4	5	6	5	4	3	2	1	0

In the above table the sequence of crop- sets written in roman numbers and, for each calendar month correspondent value of displacement d is given. Observe the main diagonal, all the values of d= 0 i.e., for crop-set I, $d = 0$ if it is sowing in the month of December, for crop-set II, $d = 0$ if it is sowing in November, and so on. Consider the crop-set I particularly, it has maximum displacement d= 6, if it is sowing in the month of June, and value of d is decreases by 1, if we move one step from June i.e., for crop-set I: $d = 5$ for July and May, $d = 4$ for August and April, d= 3 for September and March, $d = 2$ for October and February and $d = 1$ for November and January. Reader may verify the table for other crop-sets by their own.

The above table shows, that maximum possible displacement is six i.e., $0 \leq d \leq 6$, we can put $d = (0,1,2,3,4,5,6)$ in equation (5.4.4), and then, values of P_m given as follows.

Table 5.4.2- Representing Points of Month

	Pm
d=0	48.1481
d=1	44.4444
d=2	40.7407
d=3	37.0370
d=4	33.3333
d=5	29.6296
d=6	25.9259

5.4.3 Point of Fertilization factor (P_f)

In section 5.4.1, we have defined maximum weight of fertilization factor $F_w = 8$, i.e., $F_w = (0,1,2,3,4,5,6,7,8)$ we can calculate point of fertilization factor P_f, by multiplying each value of F_w by multiplicative constant Ω.

$$P_f = (F_w \times \Omega) \qquad \text{.........(5.4.6)}$$

For each possible value of F_w, P_f may be given as following table

Table 5.4.3- Representing Points of Fertilization

	$P_{f=}$
$F_w = 0$	0
$F_w = 1$	3.7037
$F_w = 2$	7.4074
$F_w = 3$	11.1111
$F_w = 4$	14.8148
$F_w = 5$	18.5185
$F_w = 6$	22.2222
$F_w = 7$	25.9259
$F_w = 8$	29.6296

5.4.4 Point of Irrigation factor (Pi)

From the section 5.4.1, possible values of irrigation factor I_w is equal to 3,2 or 1 for high, medium and low irrigation respectively. Then irrigation factor P_i:

$P_i = (I_w \times \Omega)$ .... (5.4.7)

For each possible value of I_w, P_i is given in following table

Table: 5.4.4- Representing Points of Irrigation

	$P_{i=}$
Iw = 1	3.7037
Iw = 2	7.4074
Iw = 3	11.1111

5.4.5 Point of Seed-Quality factor (Ps)

Similarly, possible values of seed-quality factor S_w is equal to 3, 2 or 1 corresponds to high , medium and low quality of seed. Then seed-quality factor:

$P_s = (S_w \times \Omega)$ ....(5.4.8)

For each possible value of I_w , P_i is given in following table

Table: 5.4.5- Representing Points of Seed Quality

	$P_{s=}$
Sw = 1	3.7037
Sw = 2	7.4074
Sw = 3	11.1111

5.5 Calculating Decision Coefficient 'α'

To calculate decision coefficient α for any particular case, first we calculate decision coefficient 'α_{max}' for ideal state, relook the equation 5.4.2 and equation 5.4.3

$$T_w = M_w + F_w + I_w + S_w \quad \text{...(5.4.2)}$$

$$T_w = 13 + 8 + 3 + 3 = 27 \quad \text{...(5.4.3)}$$

We now find decision coefficient 'α_{max}' for ideal state by multiplying each coefficient of equation 5.4.3 with Ω

$$\alpha_{max} = (13 + 8 + 3 + 3) \times \Omega$$

$$= \mathbf{27 \times \Omega}$$

$$= 27 \times \mathbf{3.7037 = 99.9999}$$

For a particular case the value of decision coefficient 'α', can be calculated by adding values of P_m, P_f , P_i and P_s i.e

$$\boldsymbol{\alpha} = P_m + P_f + P_i + P_s \quad \text{... (5.4.9)}$$

5.5.1 Experimental Results

We are including some experimental results for demonstration purpose of our model; the procedure of calculating decision coefficient 'α' is also included for different cases. We have taken examples to simulate both of the major factors seeding time and, fertilization factors, further we have added some cases to visualize the effects of irrigation factor and seed quality facto. Consider following case that shows mathematical results regarding performance of crop.

5.5.1.1 Example (Case-1)

Let any crop of crop-set I, say - wheat is seeded in March beside its standard seeding month December in soil type 2 i.e., 20 % Urea, 30% Phosphorus, 10% Iron, 10% Nitrogen and 10% Zinc since the ideal soil type for crop-set is *S2* therefore it gets maximum points of fertilizer factor, $F_w = 8$. For the simplicity it is also assumed that the quality of seed is up to the standard level and irrigation is up to the mark. Then

Crop set is = I i.e. $C_k = 1$

Month Sequence is March i.e., $M_j = 3$

Then $P_m = (M_j + C_k) = 3 + 1 = 4$, from the algorithm 4.2.1 we find the value of displacement i.e., d = 3, however, more easily go through the table and you can find the same value of d for the crop-set I in the month of November, putting d = 3 in equation 5.4.5 we get:

$P_m = 48.1481 - (3 \times 3.7037) = 37.0370$

As we have assumed that the availability of fertilizers in soil type is in ideal state, then points for fertilizer factor are maximum i.e. $F_w = 8$ since from equation 5.4.6 value of P_f

$P_f = 8 \times 3.7037 = 29.6296$

By the example the weights for irrigation $I_w = 3$ and seed quality $S_w = 3$, then from equation 5.4.7 and 5.4.8, $P_i = 11.1111$ and $P_s = 11.1111$. We can calculate the value of decision coefficient **'α' by equation 5.4.9 i.e.**

$$\alpha = \mathbf{37.0370} + 29.6296 + 11.1111 + 11.1111$$

$$= 88.8888\ \%$$

5.5.1.2 Example (Case-2)

Let any crop of crop-set I, say - wheat is seeded in January in soil type-4 i.e., 30% Phosphorus, 20% Nitrogen, 10 % Calcium and 20% Potassium, however from the model the ideal soli type for crop-set I is S2. Therefore, points of fertilizer factor, $F_w < 8$. In this example we assume that the quality of seed is medium level i.e., $S_w = 2$, but the value and irrigation factor is low $I_w = 1$

Then according to the model

Crop –set is = I, i.e. $C_k = 1$

Month Sequence is January i.e., $M_j = 1$

Then $P_m = (M_j + C_k) = 1 + 1 = 2$, from the algorithm 5.4.2.1 we find the value of displacement i.e., d = 1, putting d = 1 in equation 5.4.5 we get

$P_m = 48.1481 - (1 \times 3.7037) = 44.4444$

Ideal Land type for crop set I is S2 that has i.e., 20 % Urea, 30% Phosphorus, 10% Iron, 10% Nitrogen and 10% Zinc, in this case the crop is seeded in soil type S4 i.e., 30% Phosphorus, 20% Nitrogen, 10 % Calcium and 20% Potassium. Then total point of fertilization may me calculated by finding the difference of fertilizers availability in *S2* and *S4*.

Required Fertilizers (L2) =U2 + P3 + Fe1 + N1 + Z1
Available Fertilizers (S4) = 0 + P3 + 0 + N2 + 0 + Ca1 + K2

Total Points F_w = 0 + 3 + 0 + 1 + 0 + 0 + 0 = 4

Then from equation 5.4.6 value of $P_{f'}$ is gives as

$P_f = 4 \times 3.7037 = 14.8148$

By the example the weights for irrigation $I_w = 1$ and seed quality $S_w = 2$, then from equation 5.4.7 and 5.4.8

$P_i = 3.7037$ and $P_s = 7.4074$. We can calculate the value of decision coefficient 'α' by equation 5.4.9

α = **44.4444+ 14.8148** + 3.7037 +7.4074

= 70.3703 %

5.5.1.3 Example (Case-3)

For our third example we combining the situation of both of the above case and assuming the crop-set I, is seeded in month of March in the soil type 4. This time we consider the irrigation factor is up to the mark but seed quality is of medium quality. Then according to the model

Crop set is wheat i.e. $C_k = 1$

Month Sequence is March i.e., $M_j = 3$

Then $P_m = (M_j + C_k) = 3 + 1 = 4$, from the algorithm 5.3.2.1 we find the value of displacement i.e., $d = 3$, putting $d = 3$ in equation 5.4.5 we get

$P_m = 48.1481 - (3 \times 3.7037) = 37.0370$

Again, total point of fertilization may me calculated by finding the difference of fertilizers in S2 and S4.

Required Fertilizers (S2) = U2 + P3 + Fe1 + N1 + Z1
Available Fertilizers (S4) = 0 + P3 + 0 + N2 + 0 + Ca1 + K2

Total Points $F_w = 0 + 3 + 0 + 1 + 0 + 0 + 0 = 4$

Then from equation 5.4.6 value of P_f, is gives as

$P_f = 4 \times 3.7037 = 14.8148$

This time the weights for irrigation I_w = 3 and seed quality S_w = 2, then from equation 5.4.7 and 5.4.8

P_i = 11.1111 and P_s = 7.4074. We can calculate the value of decision coefficient 'α' **by equation 5.4.9**

α = **37.0370** + 14.8148+ 11.1111 +7.4074

= 62.9629 %

5.6 Software Demonstrating for Computation

We are demonstrating software to calculate the values of decision coefficient α for any crop-set at any seeding month and for any type of soil. Therefore, all the possible inputs of above mentioned agricultural attributes may be provided to the software and we can find the values of decision coefficient 'α'.

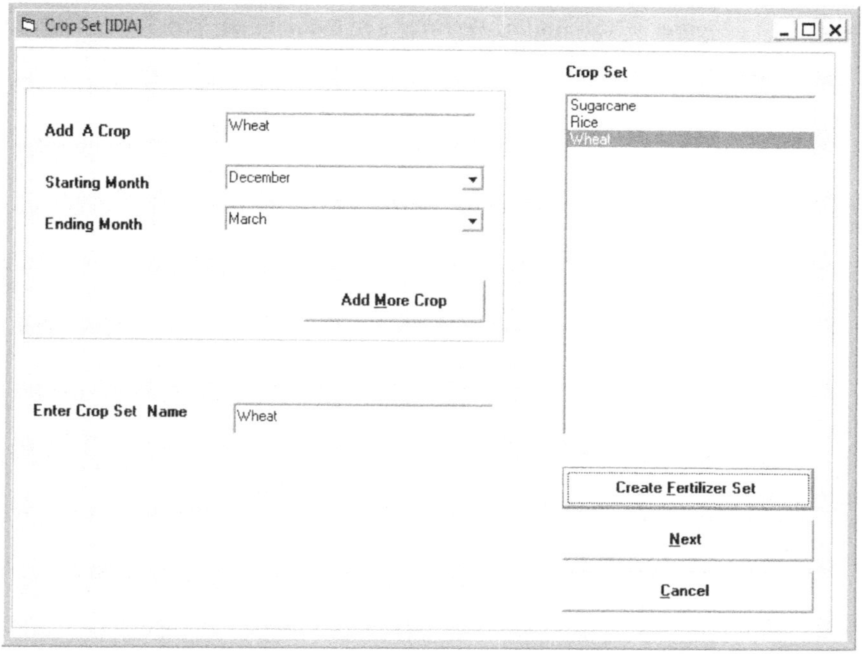

Figure: 5.6 (a)-Adding Crop into Crop set using Software Module

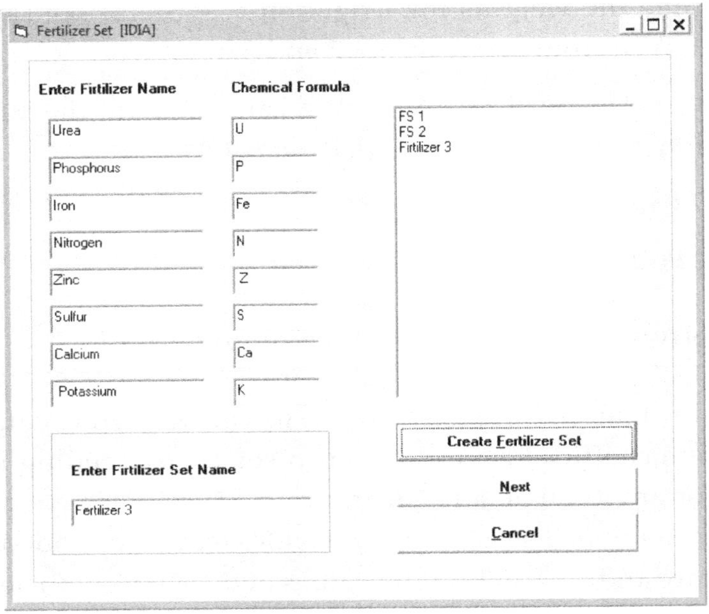

Figure: 5.6 (b)- Adding Fertilizers to build a Fertilizer Set

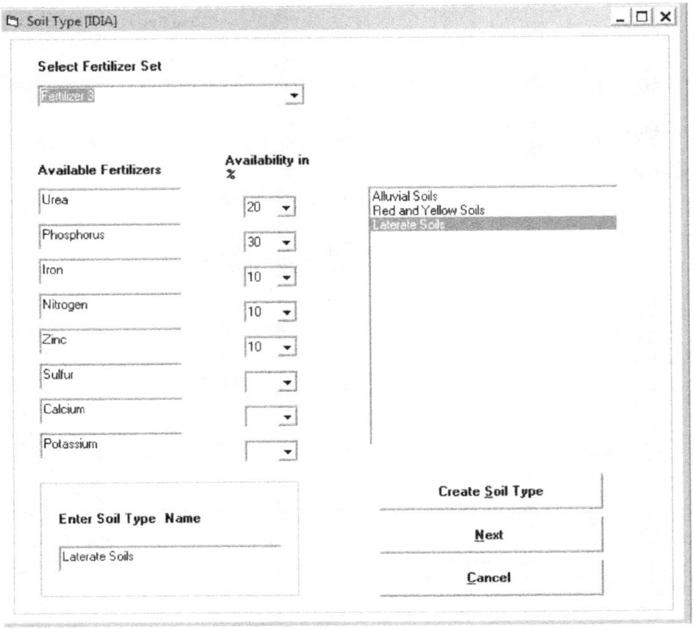

Figure: 5.6 (c)- Defining Soil Type

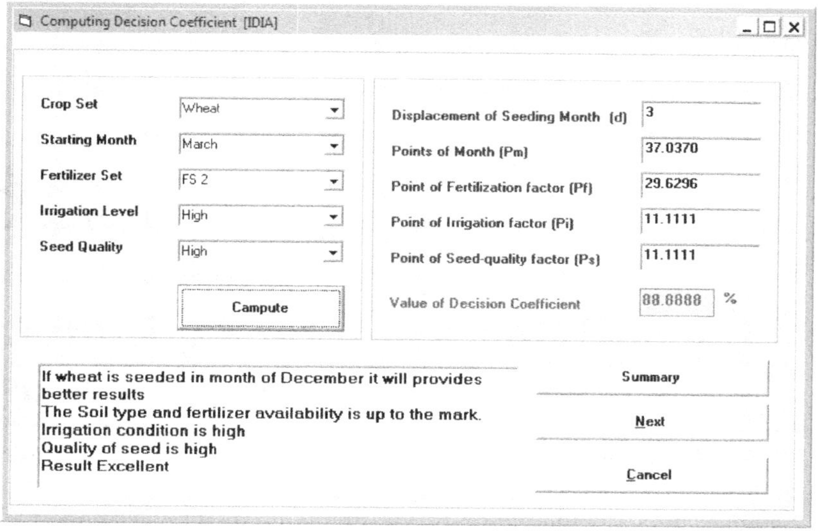

Figure: 5.6 (d)-Computing Decision Coefficient 'a'

5.7 Chapter Conclusion

The main contribution of this chapter lies in the preparation of framework on agricultural data warehouse. The study includes identification and classifying of major agricultural attributes and places them in data warehouse schema. The model provides integrated information about whole crop cycle for the best results. We have also described some novel case of change such as starting time of crop, percentage of fertilizers and impact of irrigation and seed quality. The model is supported by mathematical equations, and success percentage of crop in a particular weather and land type can easily be evaluated using these equations. This model may consider as a framework of agricultural decision support system and response to the ad-hoc queries of farmers using data mining techniques.

Chapter 6

Data Mining Agent for Agricultural Data Warehouse

Agent technology and data mining interaction is an emerging area of Information Technology. Autonomous Data Mining Software is a good idea to retrieve useful information from extremely large data warehouses. In this article we present a multi- functional data mining agent to assist farmers and economist for decision making. Agricultural data warehouse collects information from lots of information sources. The information stored in data warehouse is huge and complex in nature. There are many data mining tools and algorithm that are used for purpose of information retrieval from such large data warehouses. Multi agent systems are tools that coordinate the retrieval, management, selection and manipulation of information. They find application ground in situations where the amount of information available for a user or the solution to a problem is unmanageably large and some way to filter it down to a selection of the most essential and useful pieces are needed.

We have chosen a multi-agent approach to enable continuous updating of data warehouse views as transactions are executed at the sources. Various Multi-agent models have been developed; C. C. Henry Chan, Chi Bin Cheng, Chih Hsiung Hsu (Chan, Cheng, Hsu, 2007) introduced an autonomous agent that represents the owner of an online store to bargain with customers. This multi-agent approach applied to develop a decision support system to mine data from agriculture data ware house.

6.1 RDBMS based DSS and Data Warehouses

Relational Database Management Systems (RDBMS) are widely used by many operational databases the operational transaction made for maintaining data that documents everyday operations. The applications that update such operational data or transactions make typically small changes and a large number of transactions. RDBMS have been optimised to perform reliably and efficiently those operations, usually named online transaction processing (OLTP) applications (i.e., applications that support many users executing small transactions). Current and historical data has also to be comprehensively analysed and explored, identifying useful trends and creating summaries, to support decision making.

Data warehouses are updated continuously when the operational system performs a transaction. The Data warehouse then generates transactions which are passed back to the operational system. Data in the *Data warehouse* is regularly updated from the Operational Database. The data in *Data warehouse* is mapped and transformed to meet the data warehouse objectives. Traditional database supports on-line transaction processing (OLTP), which includes insertions, update, and deletions of data and also support information query requirements. Traditional relational database is optimized to process queries that only touches a small part of the database, the transactions deal with insertions or updates of few tuples per relation to process. They cannot be optimized for OLAP, DSS, or data mining. By contrast, data warehouse is designed precisely to support efficient extraction, processing, and presentation for analytics and decision-making purposes. OLAP systems are designed for use by data scientists, business analysts, and knowledge workers, and they support business intelligence (BI), data mining, and other decision support applications. OLTP is optimized for executing online database transactions while the OLTP systems are designed for use by frontline workers (e.g., cashiers, bank tellers, part desk clerks) or for customer self-service applications (e.g., online banking, e-commerce, travel reservations)

In comparison to traditional databases, data warehouses generally contain very large amounts of data from multiple sources that includes databases from different data models and sometimes files, acquired from independent systems and platforms. To make data mining more efficient the data warehouse should have summarized collection of data. In fact, for very large database running in terabytes of data, successful use of information retrieval and data mining will depend first on the construction of data warehouse. The overall process of designing data warehousing and data mining is given by Ramez Elmasri and Shamkant B. Navathe "Fundamentals of Database Systems", Addisson-Wesley, 2000.

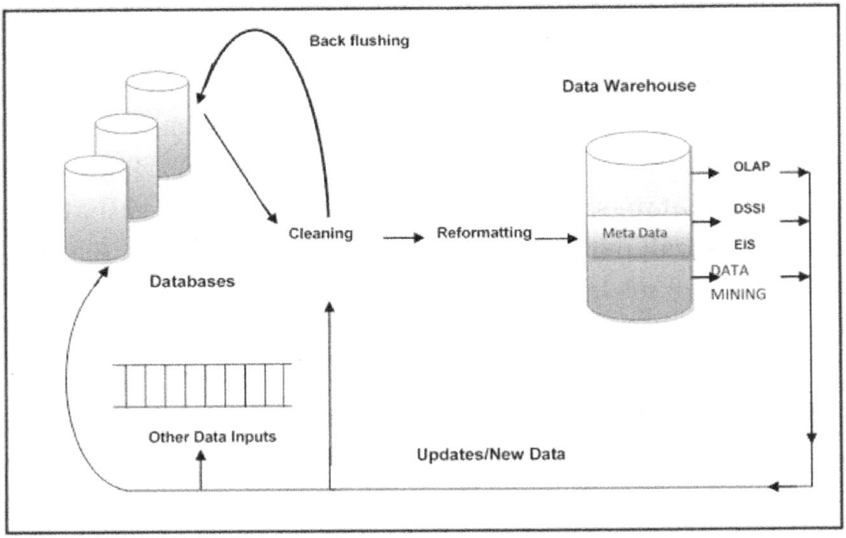

Figure: 6.1- Overall Processes of Data Warehousing

6.1.1 Self Maintainability of Data Warehouse

Data warehouses store materialized views in order to provide fast access to information that is integrated from several distributed data sources [DEB95]. The sources of agricultural data are heterogeneous in nature and collected from diversified field and formats. accordingly, maintaining a materialized view in a data warehouse is different from the maintenance of traditional view of data warehouse where the view and the data are stored in the same database. In particular, when changes are reported by one data source it may be necessary to access base data from other data sources in order to maintain the view [HGW+ 95].

A data warehouse organizes and stores consolidated data derived from distributed, autonomous data sources for OLAP (Online Analytical Processing) and data mining analyses. The data warehouse views must be kept up-to-date as changes are made at the sources. The data warehouse functioning not only involves storage and retrieval of data, but feed-backing of data and self-learning in also an important issue. Commercial systems typically designate view maintenance down time when the data warehouse is disabled for data updates. The maintenance time could range from several hours to several days, depending on the size of the data warehouse, the data sources, and the volume of updates. The self-maintenance in data warehouse can be archived through maintaining the views and materializing supplementary data. The incremental update mechanisms are most suitable for self-maintainability in large and complex data warehouses. The other issues related to the data warehouse view maintenance are: algebraic approaches, the concurrency control problem, database rule related problems. However, many studies, concluded that the differential method is superior to the others. This study advocates use of feed-backing approach for availing self-maintainability in the proposed agricultural data warehouse as well as similar complex data warehouses.

6.1.1.1 Feed backing and Machine Learning

This s model follows a feed backing approach of data to support machine learning. Machine learning is often used for prediction or classification of knowledge, which applied fuzzy logic in data mining and information retrieval. With Machine Learning, the computer makes a prediction and then, gives feedback. It learns through examples, domain knowledge and feedback. When a similar situation arises in the future, this feedback is used to make the same prediction or to make a completely different prediction.

This predictive modelling approach has two phases. During the learning phase a sample of database is use to train the system properly, while in training phase, the historical or sampled data are used to create a model that represents those data. It is assumed that this modal is representative not only for sampled data but also for database as a whole and for future data as well.

6.2 Agent Technology

Agent technology and data mining interaction is an emerging area of Information Technology. The Agent-based data mining and warehousing is presently a field of the Invention, that relates to data storage and retrieval and more particularly to an agent-based networking system for creating a data warehouse and to apply methods of accessing the required information for real time retrieval of purpose specific data.

The typical approach to data mining is to start with designing of data warehouse, that is, started with creating a catalogue of database attributes, selecting the schema, and removing ambiguities in gathered information. The design of a data warehouse is concerned with describing dimensions of multidimensional schemes and methods of integrating other technologies so that they can be accessed through manageable framework. This involves data storage, data selection, data cleaning and an infrastructure for updating databases once new knowledge or representations are developed. Autonomous Data Mining Software is a good idea to retrieve useful information from extremely large data warehouses. We present a prototype based on multi- functional data mining agent, that will be supposed to assist farmers and economist for decision making. Agricultural data warehouse collects information from lots of information sources. The information stored in data warehouse is huge and complex in nature. There are many data mining tools and algorithm that are used for purpose of information retrieval from such large data warehouses. We are proposing data mining agent operates on agricultural data warehouse and discovering information that is used to make decision. We have also proposing a model to develop multi-attribute-based data mining agent.

6.2.1 Agent Mining

The term Agent Mining refers to the use of intelligent agents in the field of data mining in order to achieve automaticity and to enhance the knowledge discovery by providing high performance and scalability. Due to the autonomous, flexible, adaptable and rational nature, agents are suitable choice for complex, multisource, and distributed mining. Developing Multi-agent-based data mining agent is an excellent idea that allows the system to sense nature of the ad-hoc query and select the particular data mining technique that is most preferable and suitable to the ad-query made by decision makers. L. Cao, V. Gorodetsky & P. A. Mitkas have well explained data mining for agent in his work *"Agent Mining: The Synergy of Agents and Data Mining," IEEE Intelligent Systems, pp. 64-72, May/June 2009.* that states bellow:

In data mining for agents, data mining is used to extract knowledge from large datasets in the form of decision trees or data induces rules, which provide logic for the intelligent agents.

- *The agent-data mining collaboration may occur and can be analysed in a number of diverse dimensions: Resource dimension at data, information, and knowledge levels.*
- *Infrastructure dimension at infrastructure, architecture, and process levels.*
- *Learning dimension at learning methods, learning capabilities, and performance levels.*
- *Interaction dimension for coordination, cooperation, negotiation, and communication. Social dimension in social and organizational factors for instance, in human roles.*
- *Performance dimension in the performance enhancement of one end of the coupled system.*
- *Interface dimension at the human-system interface, user modelling and interface design level.*
- *Application dimension in applications and domain problems.*

–(L. Cao, V. Gorodetsky, P. A. Mitkas)

Further, a new dimension is added by, A. M. Al-Barky and J. Ali in his work *"Intelligent mining agent," in 8th International Conference on Computing Technology and Information Management (ICCM), 2012, vol. 1, p. 23"* that states:

Agent mining provides for the automation of the mining steps enabling non-experts to use the system while assisting the work of experts too. Apart from providing high performance, the data mining process supported by agents helps to increase the quality of knowledge obtained, simplify the process of identifying patterns from huge data volumes as well as help in take good decisions in real time.

–A. M. Al-Barky and J. Ali

6.3 Data Mining Agent for Agricultural Data Warehouse

Agricultural data warehouse has consolidated information from several sources of agricultural practices and small databases. We have given a multidimensional data warehouse model in chapter-4 to store this information and consolidate it before placing into the warehouse. The data received from many heterogeneous sources is required a materialized view. We have defined several dimensional in to the multidimensional data cube to store this data-sets. The materialized view of data warehouse is created, managed and used by specialized Multi Agent System (MAS).

The RDBMS oriented SQL queries are frequently embedded in a host language programs. Many queries involve conditions over time or require aggregation over time periods. The traditional SQL provides poor support for such time series analysis, but with the help of autonomous MAS this can be improve significantly. On the other hand, the RDBMS also has no way to recognize and exploit optimisation opportunities arising from executing many related queries together. Here, agent technology may provide special support in order to help users. Three broad classes of analysis tools are available:

- Some systems support a class of stylised queries that involve group-by and aggregation operators and provide support for complex Boolean conditions, statistical functions and features for time series analysis. These applications are called Online Analytic Processing (OLAP). In these systems, the data is thought as a multidimensional array and is influenced by end-user tools.

- Some RDBMS that supports traditional SQL-style queries are also designed to support OLAP queries efficiently. Such systems stand for RDBMS optimised for decision support applications. The proposed integrated data warehouse also supports OLAP and OLTP for mining of data.
- The use of analysis tools is also applicable in order to find interesting or unexpected trends and patterns in large data sets, in addition to the built-in and pre-listed queries. The goal of knowledge extraction is to support exploratory analysis over very large data sets. Evaluating OLAP or knowledge extraction queries over globally distributed data is likely to be slow. The natural solution is to create a centralized repository of data warehouse.

6.3.1 Multi Agent Approach

Multi-agent Systems (MAS) set a useful paradigm in problem-solving, via theorem proving. The agent-based computing has been hailed as a significant breakthrough in problem solving and as a new revolution in software development and analysis. The agents are the focus of intense interest on many sub-fields of Computer Science. The MAS are being used in a wide variety of applications, ranging from small systems to large, open, complex and critical ones i.e., agents are not only a very promising technology, but are emerging as a new way of thinking. A conceptual paradigm for analysing problems and for designing systems, for dealing with complexity, distribution and interactivity, may be a new form of computing and intelligence.

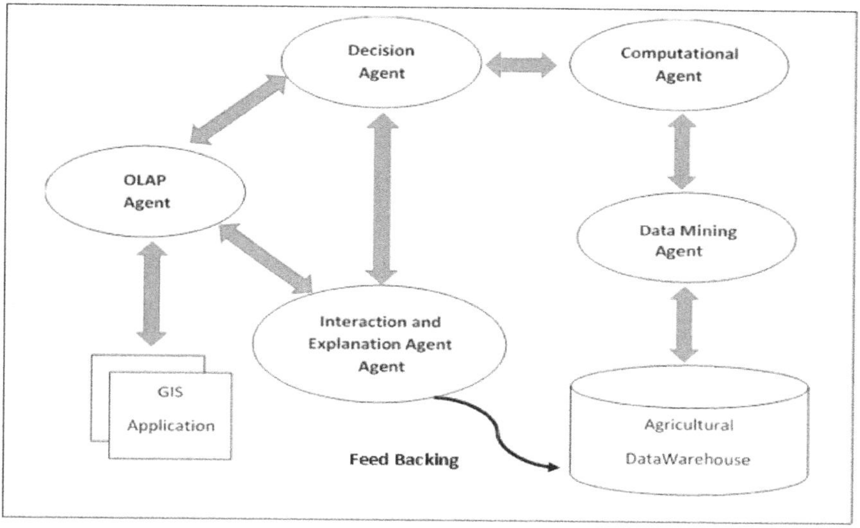

Figure: 6.3.1- Multi Agent System for Data Mining

The Data mining process in agricultural data warehouse is a complex task. Because it carries variety of functions are involved in it. We can use various types of agents to store update and mining of heterogeneous data-set, stored in multidimensional schema of agricultural data warehouses. The use of agent-based technologies for information retrieval, and information extraction is being discussed since early 90's (e.g. [PLS92, JP98, PTU99]).

An agent-based data mining and process for building an agricultural data warehouse required application specific information. The agriculture system required various functional and resource agents to communicate with agricultural data warehouse in order to collect relevant data. We are explaining the prototype model for "*MAS based Data Mining Application*" that is similar to the Broker's Lounge platform developed in

GMD-FIT. This study includes 5-types of agents based on their function.

6.3.1.1 Data Mining Agent

The Data mining agent is a processing agent that is used to retrieve information from data warehouse. It acts like mining automated that mines meaningful information against user's query. It also removes the un-useful information from so that useful information could be used properly. The data mining agent is an automated and intelligent program designed to find out the mining pattern through analyses of frequent users' queries. It also identifies patterns among the data supposed to be retrieved and can record the change in query pattern with time. These agents are typically used to detect the trends in data requirements, and to predict future queries of the user. In addition to recording patterns, data mining agents also is also responsible for selecting appropriate data mining tool for particular type of query.

Developing a data mining agent for agricultural data warehouse is an automated approach to apply most suitable data mining tool for a specific type of problems. In effect, this agent will reduce human interaction while selecting a statistical or data mining tool for retrieve information. This agent is programmed to monitor nature of the query and to filter a particular data mining technique to apply on the data stored in warehouse. In this way, a data mining agent acts to save valuable human time, as it is no longer required to engage a human for it. Additionally, data mining agents can be used to sift through database records, retrieving specific requested information that would otherwise prove tedious or difficult for a human to retrieve. For example, a data mining agent can easily and tirelessly sift through millions of records whereas a human could become tired and make mistakes during a particularly long search, an agent will never fail to retrieve its stated objective.

6.3.1.2 Computational Agent

The Computational agent is an automatic processing module that is used to compute the data retrieved from data mining agent. It is basically doing the statistical analysis on the data outputted from data mining agent. In order to detect any pattern or association between attributes it also applies perform some statistical computation on the data. Another purpose of this agent to compute the decision coefficient (as explained in chapter 5) for a particular crop-set in a particular soil type and climate conditions. As explain in chapter 5 the computation of decision coefficient is the core of this work. Therefore, the Computational Agent must be design carefully ad optimally. After computation the computed values are stored in the multidimensional schema that reduce the processing overhead during farmers; ad-hoc queries.

6.3.1.3 Decision Agent

This is the agent that needs to make decision based on the values provided by computational agent. It is an automatic processing module that receive the computed values from computational agent and after processing provide a concrete decision about selection crop for a particular crop according to computed decision coefficient. The Decision agent also provides prediction of crop production, suggest improvement areas, and also provides concluding remarks to the farmers queries particularly those that related to crop productions. Another important work of decision agent is to co-ordinate between other agents. Therefore, it collects and use information like what is available in database, what values are computed, and what is need to be analysed by the user. Decision agent also make decision what type of data is ready to be display through OLAP or what type of data needs explanation or human interaction before visualization.

6.3.1.4 Interaction and Explanation Agent

The explanation agent works as a mediator between OLAP agent and decision agent. This agent coordinates between what is asked and what is available in to information reservoir. This also provide "what if" analysis to the user. The interaction agent is majorly responsible for providing explanation of the data provided by decision agent computation agent. It is similar to converting the mathematical and statistical data in to language interpretation that can be easily understand by human user. It is observed that the user wants to know the appropriate cause of the problem and also want to know the solution or alternatives of the problems. Generally, user do not have interest to know the details of statistical analysis or complex mathematical computation. Keeping this in mind the one purpose of interaction and explanation agent is to explain the problem and their solution without involving mathematical details or complex reasoning of decision-making process.

6.3.1.5 OLAP Agent

The agricultural data warehouse is large and complex and it required to handle a scalable OLAP framework. The high volume of agro data and flow rate of data pose serious scalability and performance challenges. In order to overcome these issues and also to automate the whole operation of data warehousing (that includes capturing of data, cleaning, summarizing, and loading) an OLTP agent is recommended. The automation of OLTP function is intended to increase performance while storing and mining large set of statistical records in data warehouses. Like other business intelligence applications agricultural data warehouse also require continuous collection of transactions in its repository. On the other hand, the mining operation also needs continuous analysis of records for feed-backing the records in data warehouse. As discussed above the decision agent gives concluding remark in natural language, these remark needs to process, compile and feeded-back to the warehouse. The proposed OLAP agent also interprets the meaning of such remarks and apply efficient filtering of useful data before feedback to warehouse.

Another major responsibility of OLAP agent is to provided assistance to the visualization process. In this study, the OLAP agent is supposed to provide answers of the query in interactive and suitable way. The GIS applications have been working to visualize various geographical facts such as land-type, rainfall, crop details etc. This OLAP agent will also deal with GIS tools application for efficient visualizing using maps and satellite linked online images.

6.4 Chapter Conclusion

A review of the reported research work on multi-agent-system is included in the early section of the chapter. After study and review, some important issues have been identified. Further, the chapter focuses on the advantages of using agent technology for data warehousing and data mining. It also includes a detail framework for developing a multi-agent software (MAS) in order to retrieve information from complex warehouse. A detail framework to design a data mining agent for agriculture data warehouse is discussed in the chapter. Another important issue discussed is concept of predictive data mining approach that supports in making automatic decision, while sending data to OLAP or to feedback to the warehouse repository. The use of multi agent software for self-learning and predictive data mining are also discussed in the chapter.

Chapter 7

Summary and Conclusion

In this study, efforts have been made to provide a conceptual design of data warehouse for Indian agriculture. The proposed data warehouse implemented multidimensional schema to place agricultural parameters selected for this study. This work also led to design such data warehouse that act as decision support system, based on OLAP and data mining techniques and therefore it will also provide response to farmers' ad-hoc quires. In this work an efficient computational model has been introduced that helps to stores ready-to-use values in the data warehouse. The given computational model is well conceptualized and described for computing decision coefficient. The value of decision coefficient is further stored in the multidimensional data cube. The decision coefficient helps to improve the performance of data warehouse. It also identifies various needs and lacks in cropping system.

For optimal data mining purpose, the multi-agent based approach is suggested that also helps to automate the DSS. The agent based data mining also incorporates the self-learning abilities into data warehouse. It keeps data warehouse keep update with time.

This chapter mainly discussed the procedure, technologies and concept used for designing integrated data warehouse for Indian agriculture. It provides chapter-wise proceedings of the work, and also includes major findings, main contribution and, possibilities of future work in this area.

7.1 Concluding Remarks

The summary of this work can be understood by analysing contribution of each individual chapter. This section is to focuses the work carried out in each individual chapter. These concluding remarks focuses various methodologies, procedure, and concept used for designing an Integrated Data Warehouse for Indian Agriculture.

- The very first chapter presents a brief summary of agricultural development in India, before and after independence. The chapter throws light on the agriculture practices in India, since beginning of the history. The various initiatives taken by government of India in the field of agriculture research, and some reported work on agricultural database and information system, has also been presented in this chapter. The chapter finally presents the background of the work, the reason, motivation and objectives of this work.
- Chapter-2 provides a literature review and the survey of research works done by many researchers in agriculture sector. In this chapter we look through the existing technologies and concepts, that are likely important for conceptualization of this work. The chapter provides detail descriptions about Indian agriculture system. The existing methodology used for developing a data warehouse and DSS for agriculture are also discussed thoroughly in this chapter. The definitions salient features, functions, procedure and methodology of various concepts are explained clearly. Understanding of related technologies like DSS, MAS, and GIS are also covered in the chapter.

- Chapter-3 presents basic concepts which pertain to describe agricultural attributes. After literature review, we have found some major technologies and practices are need to define and formularize. The chapter contains functional details, procedures and methodology of various concepts used to build a data warehouse. The chapter helps to understand to identify major attributes of agriculture system. The chapter focuses on theoretical and practical methodologies of Data Warehousing such as Data Modeling, OLAP Cube Design, Query Optimization etc. This chapter manly covers the methods and tools used in data mining for large databases. The Data Mining task along with data mining techniques, functions and methods, are well discussed in the chapter. The other useful concepts related to MAS are also discussed in the chapter.

- The chapter-4 provides the detail framework of designing a multi-dimensional schema for agricultural data warehouse. The concept of dimensional modelling is applied to conceptualize dimensional of OLAP cube. Then the various dimensional namely Crop-sets, Calendar-Month and Land-Type are identified and placed in OLAP cube of data warehouse cube. Further a correspondent star schema for OLAP cube is presented with suitable graphics. Further, the primary keys are identified to connect each of the dimensions with central fact table of the star schema.

- The next important work of this chapter is the extension of star schema in to snowflake schema. The hierarchical aggregation is applied to extend each dimension to incorporate further attributes of agriculture. In snowflake schema, the microscopic view of attributes, affecting each dimension of the cube is included and explained. Some OLAP operations that are usually applied to all the warehouses are presented to retrieve the microscopic facts stored in the multidimensional database. The fact table of data cube is explained with the help of diagram and a similar schema is also written. The chapter also includes sample queries to demonstrate drill-down and rollup operations performed on OLAP cube.

- The real strength of this chapter lies in the adoption of predictive data mining techniques. Various data mining tools and algorithm may be applied to retrieve hidden information and pattern from the data store in OLAP cube, which probably responds to more complex queries. For example, a farmer might discover that a particular crop produces better results, when it is irrigated at a particular time, under a particular weather condition.
- This chapter finally provides a plot to develop A Computational Model for Agricultural Decision Support System. That can significantly estimate production of crop in a particular land type and weather condition. The same is discussed in the chapter ahead.
- The Chapter-5 contributes in the preparation of framework on agricultural data warehouse. In this chapter identification and classifying of major agricultural attributes has been made. These attributes are then placed in data warehouse schema. The efforts have been made, (i) to explore types of soil, (depends on combinatory percentage of various fertilizers), (ii) to classify crops in twelve major classes and (ii) to propose a classifier for weather based on calendar months. Further chapter includes a procedure of placing these attributes in the computational model

and systematically evaluates performance of crops for a particular soil, weather and for other agricultural attributes.

- The proposed model in the chapter-5, produces mathematical values of decision-making coefficient for production of various types of crop, in a particular land type and in a particular environmental attribute. This chapter described some novel case of change such as starting time of crop, percentage of fertilizers and impact of irrigation and seed quality. The software implementation of the model is also demonstrated and some screen snapshots are included to understand the working of the model.
- The same chapter also provides algorithm and tables to conceptualize the model in better way. The required algorithms and tables are presented for essay understanding of the reader. The graphical representation and functional details along with required mathematical procedure explained with suitable examples. Finally, some experimental results and cases are given to understand the working of the model.
- The chapter-6 presents the implementation of data mining agents in for agricultural data warehouses. Chapter gives the detail working process of multi agent software (MAS) to retrieve information from warehouse. A review of the reported research work on multi agent-based negotiation approaches has been also included. From the performed review, some important open issues have been identified and observations have been taken in this chapter.
- A detail frame work to design a data mining agent for agriculture data warehouse is describe in this chapter. Another important issue discussed in this chapter is concept of predictive data mining approach. The proposed approach

is adapted to develop an agent for decision making. The applicability of software agent for self-learning and predictive data mining are also discussed in the chapter.

- This chapter-7 is written having aim to provide a glance of the work to the readers and to discuss main contribution of the work. The additional objective of the current chapter is to provide future tracks to the researchers, that are emitted from this study. The limitations, issues, ideas and future works that are beyond the scope of this study are also discussed.

7.2 Main Contribution

This study has mainly presented a comprehensive framework to design an Integrated Data warehouse for Indian Agriculture. The study has contributed in developing a computational model for agricultural decision support system. The ultimate goal of the study is not only designing a data warehouse for agriculture but also to devolve a decision support system that could response ad-hoc queries of the farmers. This research work also provides a holistic view of agricultural attributes and help to understand the impact of these attributes on production of crop. The main contribution of the work can be explained as bellow.

- It gives complete understandings of parameters that majorly affect the crop production. It also helps to identify the untouched problem of high importance and useful in the domain of interest for research prospective in the area of data warehouse technology in agriculture sector.
- The agricultural parameters soil type, fertilizers, seeding month etc. have no standard mathematical formulation,

this research contributes to formulize these attributes using mathematical procedures and algorithms.

- The contribution of this work also includes incorporating data warehousing, DSS and agent technology in subject oriented data warehouses.
- It also presents a comparative review of already in use data warehouses and the proposed agricultural data warehouse.
- The research may become the reason of fertilizing new ideas in the mind of the researchers.

7.3 Limitations

- This work study is based on countable but major factors of agriculture system, such as Soil Type, Crop Set, and Weather Conditions etc. However, many known and unknown factors also significantly affect the result. So, there is fair chance of subjectivity error.
- This work is based on the theoretical facts of "Indian cropping system", but due to geographical changes and change in climate, the output of computational model is not error proof.
- The most adopted methodology "multidimensional schema" is used for conceptualizing this data warehouse, however other topological data structure is/are possible to store and retrieve information.
- The study is limited to provide a conceptual framework; it does not include any coding or implementation phase of software development process.
- The application of data warehousing in field of agricultural is comparatively new, so the result and performance of

the proposed model are needs to aligned with actual and practical facts.

7.4 Future Work

Over a next few year, not only will there be more efficient algorithms with better interface techniques, but also steps will be taken to develop an all-encompassing model for data warehousing. This research work may lead to allow unifying data warehouse methodology, and applying these concepts not for only agricultural data but related data warehouses of future. The computational model in this work may provide a solution to construct a data warehouse with less redundancy. The advent of relational data model and SQL were milestones in the evolution of database system. This study leads to develop universal Data Mining Query Language, which is equally useful for all programming languages and other database models. This research is likely focus on automatic aspects of warehouse that currently require significant manual intervention such as data acquisition, data quality management, and self-maintainability. The work also advocates the uses GIS applications for better GUI based software that is more essay to implement by non-technical users.

It is hoped that, this research work will give better ideas to relate information technology with Indian agriculture. This may open new paths for implementation of other technologies (such as IoT and data analysis) with the field of data warehousing. The study welcomes to the researchers to incorporate such technologies for development of more advance and Integrated Data Warehouse for Indian Agriculture.

References

[1] Abdullah, Ahsan, Brobst, Stephen, Pervaiz, Ijaz, Umar, Muhammad, Nisar, Azhar, "*Learning Dynamics of Pesticide Abuse through Data Mining*", Australasian Workshop on Data Mining and Web Intelligence, Dunedin, New Zealand (2004).

[2] Abdullah, Ahsan, Hussain, Amir, "*Data Mining a New Pilot Agriculture Extension Data Warehouse*", Journal of Research and Practice in Information Technology, Vol. 38, No. 3, 229–249, (2006).

[3] Abdul-Rahman, A. Hailes S., "*Supporting Trust in Virtual Communities*", In: Proceedings of the Hawaii's International Conference on Systems Sciences, Maui, Hawaii, (2000).

[4] Adamson Christopher, Michael Venerable, "*Data Warehouse Design Solutions*", J. Wiley & Sons, Inc., ISBN-13: 978-0471251958, (1998).

[5] Aditya K Gupta, "*Agro-Biodiversity Informatics*" using Remote Sensing and GIS", Proceedings of International Seminar "Sources of Planet Energy, Environmental and Disaster Science: Challenges & Strategies" Published by School of Management Sciences Lucknow, ISBN:978-81931392-40, (2015).

[6] Aditya K Gupta, Bireshwar Dass Mazumdar, "*Computational Model for Agricultural Decision Support System*", International

Journal of Computer Applications (ISSN No. 0975 – 8887) Volume 66, No. 15, (2013).

[7] Aditya K Gupta, Bireshwar Dass Mazumdar, *"Multidimensional Schema for Agricultural Data Warehouse"*, International Journal of Research in Engineering and Technology (ISSN No. 2319-1163) Volume 2, Issue 3, (2013).

[8] Aditya K Gupta, M.H. Khan, *"Clock Based Model for Cropping System: A Frame work for Agricultural Data Warehouse"*, National Conference, CSI, Delhi Chapter, (2007).

[9] Aggarwal, P. K. *"Global Climate Change and Indian Agriculture: Impacts, Adaptation, and Migitation"*, Indian Journal of Agricultural Sciences, (2008).

[10] Alex Berson, Stephen J. Smith, "Data Warehousing, Data Mining, and OLAP". Tata McGraw-Hill (2004).

[11] Alexander Hinesburg, Danial A. Keim. *"Clustering method for large database: From past to the future"*, Technical Report, ACM SIGMOD Tutorial, (1990).

[12] Amarnath Tripathi, A.R. Prasad, *"Agricultural Development in India since Independence: A Study on Progress, Performance, and Determinants"*, Journal of Emerging Knowledge on Emerging Markets Vol. 1 Issue. 1 (2009).

[13] Amir I., A. Nevo. *"CROPLOT: An expert system for determining the suitability of crops to plots"*, Agric. Syst., 37: 225-241, (1991).

[14] Anil K. Jain, Richard C. Dubes. *"Algorithms for Clustering Data"*. Englewood Clifts, N. J. Prentice Hall, (1988).

[15] Anil Rai, Sree Nilakanta, Kevin Scheibe, *"Indian Agricultural Data Warehouse Design"*, Conference Managing Worldwide

Operations & Communications with Information Technology, Idea Group Inc., (2007)

[16] Asner, G.P., C.A. Wessman, D.S. Schimel, and S. Archer, *"Variability in leaf and litter optical properties: Implications for BRDF model inversions using AVHRR, MODIS, and MISR"*. Remote Sensing of Environment, 63:243-257, (1998).

[17] Avison, D., Fitzgerald, G., *"Information Systems Development – Methodologies, Techniques and Tools"*, Third Edition, McGraw Hill Education: New-York, (2002).

[18] Bhavani, T.A., Ashok Gulati, Devesh Roy, *"Structure of the Indian Food Processing Industry: Towards Scaling Up and Consolidation"*, International Food Policy Research Institute, New Delhi, (2007).

[19] Biggs Stephen, *"Learning from the positive to reduce rural poverty and increase social Justice: Institutional innovations in agricultural and Natural Resources Research and Development"*, Experimental Agriculture, (2008).

[20] Blake, C. (2004), *"Doing the Knowledge"*, Proceedings of Doing the Knowledge Exploiting Knowledge Management strategies to increase quality & Profitability in building & construction, London 30th March 2004

[21] Brieman, Freidman, Olshen, Stone, *"Classification and Regression Trees"*, Wadsworth, (1984).

[22] Broner, J.P. King and A. Nevo. *"Structured induction for agricultural expert systems knowledge acquisition"*, ('omputers and Electronics in Agriculture, 87-99, Elsevier Science Publishers B.V., Amsterdam (1990).

[23] C. Westphal and T. Blaxton, *"Data Mining Solutions"*, John Wiley, (1998).

[24] C.R. Sukumar, "Precision farming may turn a reality Financial Daily" from THE HINDU group of publications, (2003).

[25] Candea C., Hu H., Iocchi L., Nardi D., Piaggio M. "*Coordination in multi-agent*" RoboCup teams. Robotics and Autonomous Systems Journal, Vol. 36, (2001).

[26] Chedad, A., Moshou, D., Aerts, J.M., Van Hirtum, A.; Ramon, H.; Berckmans, D., "*Recognition System for Pig Cough based on Probabilistic Neural Networks*". Journal of Agricultural Engineering Research 79(4): 449–457, (2001).

[27] Crist, E.P. and R.J. Kauth, "*The Tasseled Cap de-mystified. Photogram metric Engineering and Remote Sensing*", PHOTOGRAMMETRIC ENGINEERING AND REMOTE SENSING, Vol. 52, No.1, pp 81-86, (1986).

[28] David J. Hand, Heikki Mannila and Padhraic Smyth, "*Principles of Data Mining*", MIT Press, ISBN-13: 978-0262082907, (2000).

[29] Dorian Pyle, "*Data Preparation for Data Mining*", Morgan Kaufmann, (1999).

[30] E. Thomsen. "OLAP Solutions. Building Multidimensional Information", John Wiley & Sons, Inc., ISBN 10: 0471400300 / 13: 9780471400301, (2002)

[31] EDELSTEIN, HERB. "*Mining Large Databases: A Case Study*", Technology Report, Two Crows Corporation, (1999).

[32] Fischer, A., S. Louahala, P. Maisongrande, L. Kergoat, G. Dedieu. "*Satellite data for monitoring, understanding and modelling of ecosystem functioning*". In B. Walker and W. Steffen, eds. Global Change and Terrestrial Ecosystems.

Cambridge University Press: New York. pp 566-591, (1996).

[33] Ganesan V., "*Decision Support System "Crop-9-DSS" for Identified Crops*", International Journal of Computer and Systems Engineering, Vol:1, No:12, (2007).

[34] Gini Gorlinski, "*The History of Agriculture*", Britannica Educational Publishing ISBN: 9781615309214, (2013)

[35] Gio Wiederhold, "*Information systems that also predict into future*". In Lecture Notes in Computer Science, Springer-Verlag, pp.1-14, (2002).

[36] Guiteras, Raymond. "*The Impact of Climate Change on Indian Agriculture*", Department of Economics, MIT, (2007).

[37] Gulati, Ashok 2008, "*Fragmenting Bottom and Consolidating Top: India's Changing Food System and Implications for Small Holders, in India: Some aspects of Economic and Social development (eds.)*", S. Mahendra Dev and K.S. Babu. Academic Foundation, India, (2008).

[38] Gulati, Ashok, P. K. Joshi, Maurice Landes. "*Contract Farming in India: An Introduction*", (2003), available at http://www.ncap.res.in/

[39] Hal R. Varian, Lyman, Peter, "*How Much Information*", Study Report: School of Information Management and Systems at the University of California at Berkeley. (2003).

[40] Han and Kamber, "*Data Mining: Concepts and Techniques*", Morgan Kaufmann Publishers (an imprint of Elsevier), ISBN 978-0-12-381479-1, (2000).

[41] Hector Garcia-Molina, Jeffery D. Ullman and Jennifer Widom. "*Database System- The Complete Book*", Pearson Education, ISBN:9788131708422/ 813170842X (2008).

[42] Hinneburg, Daniel A. Keim "*Clustering Methods for Large Databases: From the Past to the Future*". In Alex Delis, Christos Faloutsos, Shahram Ghandeharizadeh Editors: SIGMOD 1999, Proceedings ACM SIGMOD International Conference on Management of Data, June 1-3, 1999, Philadelphia, Pennsylvania, USA. ACM Press, ISBN 1-58113-084-8, (1999).

[43] Horowitz, Sahni, Rajasekaran, "*Fundamentals of Computer Algorithms*" Galgotia Publication, ISBN-10: 9788175152571; ISBN-13: 978-8175152571 (2001).

[44] Iida, M., Umeda, M., Kaho, T., Lee, C.K. and Suguri, M., "Measurement of Annual Crops". International Conference on Precision Agriculture, St. Paul. MN. 19-22, July 1998, ASA, CSSA, and SSSA, Madison, WI, (1998).

[45] Iida, M., Umeda, M., Kaho, T., Lee, C.K. and Suguri, M., "*Measurement of Annual Crops*", International Conference on Precision Agriculture, St. Paul. MN. 19-22, July 1998, ASA, CSSA, and SSSA, Madison, WI, (1998).

[46] India's Role in World Agriculture." *Monitoring Agri-trade Policy*". European Commission, December (2007).

[47] Jain, A. K., and Dubes, R. C. "*Algorithms for Clustering Data*", Englewood Cliffs, N.J.: Prentice- Hall, (1988).

[48] K. Jain, M. N. Murty, and P. J. Flynn. "*What is data warehouse?*" ACM Computing Survey, September (1999).

[49] Kimball, R., Reeves, L., Ross, M. and Thornthwaite, W. "*The Data Warehouse Lifecycle Toolkit: Expert Methods for Designing, developing, and Deploying Data Warehouses*". John Wiley & Sons, ISBN:9781118079560, 1118079566, (1998).

[50] Kloesgen W., "*A Multipattern and Multistrategy Discovery Assistant. In Advances in Knowledge Discovery and Data Mining*", eds. U. ayyad, G. Piatetsky- Shapiro, P. Smyth, and R. Uthurusamy, 249–271. Menlo Park, Calif.: AAAI Press, (1996).

[51] Krishna, K. L., and Uma Kapila, eds. "*Readings in Indian Agriculture and Industry*". Academic Foundation, ISBN: 9788171887347, (2009).

[52] L. Silverston, W. H. Inmon, K. Graziano. "*The Data Model Resource Book*" Vol-1. J. Wiley & Sons, Inc., ISBN10 0471380237, ISBN13 9780471380238, (2001).

[53] M. Berry and G. Linoff, "*Data Mining Techniques*", John Wiley, ISBN: 978-1-118-08745-9 (1997).

[54] M. Blaschka. "*FIESTA: A Framework for Schema Evolution in Multidimensional Information Systems*". Proc. of 6th. CAISE Doctoral Consortium, Heidelberg, Germany, (1999).

[55] M. Golfarelli, Stefano Rizzi. "*A Methodological Framework for Data Warehouse Design*". Presented in DOLAP '98, ACM First International Workshop on Data Warehousing and OLAP, Bethesda, Maryland, USA, (1998).

[56] M. R. Rao, "*Legumes Production in Traditional and Improved Cropping Systems in India*", in symposium on Grain Legumes Production, November 1980, Chiang Mai, Thailand, (1982).

[57] Major, J., and Mangano, J., "*Selecting among Rules Induced from a Hurricane Database*". Journal of Intelligent Information Systems 4(1): 39–52, (1995).

[58] Margaret H. Dunhan, "*Data Mining Introductory & Advanced Topics*", Pearson Education ISBN: 9788177587852, (2003).

[59] Marotta, "*Data Warehouse Design and Maintenance through schema transformation*". Master thesis Universidad de la Republica, Uruguay, (2000).

[60] Mazumdar, B. D. Mishra, R.B. "*Cognitive Parameter Based Multiagent System for Negotiation in e-Commerce*", International Journal of Intelligent Information Processing, Vol-6 (4):46-70, (2009).

[61] Mazumdar, B. D. Mishra, R.B. "*Multiagent Paradigm for the Agent Selection and Negotiation in a B2c Process*" International Journal of Intelligent Information Technologies", Vol-5(1):61-82, (2009).

[62] Michale J.A. Berry Gorden S. Linoff, "*Data Mining Techniques for Marketing, Sales and CRM*", Wiley Publishing Inc. ISBN:9780471470649, 0471470643 (2004).

[63] Mishra Srijit and Gopikrishna SR. "*Nutrient Based Subidy (NBS) & Support Systems for Ecological fertilization in Indian Agriculture*", Policy Brief, Greenpeace, India, (2008).

[64] Mucherino, A.; Urtubia, A., "*Consistent Biclustering and Applications to Agriculture*", IbaI Conference Proceedings, Proceedings of the Industrial Conference on Data Mining (ICDM10), Workshop Data Mining in Agriculture (DMA10), Springer: 105–113, (2010).

[65] N. Haie and R.W. Irwin, "*Diagnostic expert systems for land drainage decisions*". Irrig. Drain. Syst. 2(2): 139-146, (1988).

[66] Nevo and I. Amir., "*CROPLOT: An expert system for determining the suitability of crops to plots*". Agric. Syst., 37: 225-241, (1991).

[67] O. W. Morgan, M.J. McGregor, M. Richards and K.E. Oskouri, "*SELECT: An expert system shell for selecting*

amongst decision or management alternatives". Agric. Syst., 31: 97-110, (1989).

[68] P. Faratin, C. Sierra, and N.R. Jennings, "*Using Similarity Criteria to Make Trade-Offs in Automated Negotiations,"* Artificial Intelligence, vol. 142, no. 2, pp. 205-237, (2002).

[69] Pandey, V.M. Chowdary, B.C.Mal and P.P.Dabral "*Estimation of runoff for agricultural watershed using SCS Curve Number and Geographic Information System"*, Map India Conference (2003).

[70] Panzarasa P., Jennings N.R., Norman T.J., "*Formalizing Collaborative Decision-making and Practical Reasoning in Multi-Agent System"*, vol. 12 No 1 pp. 55-117, (2002).

[71] Paulraj Ponniah, "*Data warehousing Fundamentals"*. J. Wiley & Sons, ISBN:9788126509195, 8126509198, (2006).

[72] Plant R.E., R.D. Horrocks, D.W. Grimes and L.J. Zelinski., CALEX / Cotton: "*An integrated expert system application for irrigation scheduling"*. American Society of Agricultural Engineers. 35(6): 1833 -1838, (1992).

[73] Quinn M. "*Evolving co-operative homogeneous multi-robot teams"*, Proceedings of International conference "2000 IEEE/RSJ", Takamatsu, Japan, pp.1798–1803, (2000).

[74] R. Agrawal, A. Gupta, S. Sarawagi, "Model*ing Multidimensional Databases"*, Proceedings 13th International Conference on Data Engineering, ISBN:0-8186-7807-0, (1997).

[75] R. Kennedy, Lee, Reed, and Van Roy, "*Solving Pattern Recognition Problems"*, Prentice-Hall, (1998).

[76] R. Kimball, Margy Ross, Warren Thornthwaite, Joy Mundy, Bob Becker "*The Data Warehouse Lifecycle Toolkit"*, J. Wiley & Sons, Inc. BN:9781118079560, 1118079566 (2011).

[77] R. Srinivasan, B.A. Engel and G. N. Pandyal, "*Expert system for irrigation management (ESIM)*". Agric. Syst., 36: 297-314, (1991).

[78] Rakesh Agrawal, Ramakrishnan Srikant, "*Fast algorithms for mining association rules in large databases*", Proceeding of the International Very Large Databases Conference, (1994).

[79] Ramesh, P., N. R. Panwar, A. B. Sing, S. Ramana, Sushil Kumar Yadav, Rahul Shrivastava, and A. Subba Rao. "*Status of Organic Farming in India*", General Article, Current Science, Vol-98(9) -(2010).

[80] Ramez Elmasri and Shamkant B. Navathe, "*Fundamentals of Database System*", 3rd edition Addison Wesley, ISBN: 981-405330-9, (2000).

[81] Ranjan Rajiv, and V. P. Upadhyay. "*Ecological Problems Due to Shifting Cultivation*", Current Science 77(10):1246(1999).

[82] Ravi, N. and Jagadeesha, C.J., "*Precision Agriculture, Training course on Remote Sensing and GIS Applications in Agriculture*", May 27th -7th June, 2002, RRSSC- Bangalore, pp: 225-228, (2002).

[83] Reardon, Thomas and Ashok Gulati, "*The Rise of Supermarkets and their Development Implications-the International Experience Relevant for India*" IFPRI Discussion Paper 00752, IFPRI, Washington D.C., February (2008), available at http://www.ifpri.org/pubs/dp/ifpridp00752.asp.

[84] Rohit Chandra, Aditya K Gupta, "*Scholastic Study of Impact of Software Mining in Its Modernization*", Proceedings of National Conference on "Innovations in Sciences and Technology Towards Industrial Development", jointly

published by Institute of Engineers (INDIA) C.G. State Center Raipur and Rungta College of Engineering & Technology, Bhilai, January 2008.

[85] S. Chaudhuri, U. Dayal. "*An overview of Data Warehousing and OLAP Technology*". SIGMOD Record 26(1), (1997).

[86] S. Mohan and N. Arumugam, "*Expert system applications in irrigation management: an overview*", 17: 263-280, (1997).

[87] Schatzki, T.F.; Haff, R.P.; Young, R.; Can, I.; Le, L-C.; Toyofuku, N., "*Defect Detection in Apples by Means of X-ray Imaging*", Transactions of the American Society of Agricultural Engineers 40(5): 1407–1415, (1997).

[88] Scott Spangler and Jeffrey Kreulen, "*Introduction to Mining the Talk: Unlocking the Business Value in Unstructured Information*", E Book: ISBN-10: 0-13-233953-6, ISBN-13: 978-0-13-233953-7, (2008).

[89] Sharma, Arun K. "*The Potential for Organic Farming in the Drylands of India*" Arid Lands Newsletters, (2005)

[90] Srivastava, Ravi, "*Prospects for Small Holding Agriculture in India*" presented at the workshop titled Exploring Alternatives Futures for Agricultural Knowledge, Science and Technology (KST) organized by National Council of Applied Economic Research and International Food Policy Research Institute at India Habitat Centre, Amaltas Hall, New Delhi, (2008).

[91] Subba Rao, N.S.S., "*The History of Microbiology at the Indian Agricultural Research Institute*" in: B.D. Kaushik, (Eds.): Advances in Microbiology at IARI 1961-1974. New Delhi: Indian Agricultural Research Institute pp.1-6, (2005).

[92] Tewatia, R.K., Kalwe, S.P., and Chaudhary, R.S., "*Role of Biofertilizers in Indian Agriculture*", Indian Journal of Fertilizers, V(I), p. 111-118, (2007).

[93] Trivedi, Divya, "*Food for thought*", Business Line. New Delhi, (2008), available at http://www.thehindubusinessline.com

[94] U. Fayyad, Piatetsky-Shapiro, Smyth, and Uthurusamy, "Advances in Knowledge Discovery and Data Mining", MIT Press, (1996).

[95] U.K. Shanwad, V.C. Patil, G. S. Dasog, C.P. Mansur and K. C. Shashidhar (2002) "*Global Positioning System (GPS) in Precision Agriculture*". The Asian GPS Conference, New Delhi India, (2002).

[96] U.K. Shanwad, V.C. Patil, G. S. Dasog, C.P. Mansur and K. C. Shashidhar "*Global Positioning System (GPS) in Precision Agriculture*", The Asian GPS Conference New Delhi, India (2002).

[97] V. Ganesan, "*Agricultural expert system for the diagnosis of pests and diseases*", 15th International Workshop on Artificial Intelligence in Agriculture AIA'2004, IFAC, Cairo, Egypt, pp. 107-110, (2004).

[98] Van, Veenhuizen René. "*Cities Farming for the Future Urban Agriculture for Green and Productive Cities*". Silang, Cavite, Philippines: International Institute of Rural Reconstruction, (2006).

[99] Vasant Dhar, Roger Stein, "*Seven Methods for Transforming Corporate Data into Business Intelligence*", Prentice Hall, (1997).

[100] Veenadhari, S., B. Misra and C. Singh,. "*Data mining Techniques for Predicting Crop Productivity*", International Journal of Computer Science and Technology, pp. 98-100(2011).

[101] Venkataraman, G. S., and Tilak, K.V.B.R., "*Biofertilizers in Sustainable Agriculture*", In V.Kumar, G. C. Shrotriya, and S.V. Kaore, eds. Soil Fertilty and Fertilizer Use – Vol IV. New Delhi: PR Department, Marketing Division, Indian Farmers Fertiliser Cooperative Limited (IFFCO), (1990).

[102] W. H. Inmon., "*Building the Operational Data Store*". Wiley, ISBN:9780471774235, 0471774235, (2005).

[103] W.H. Immon., "The data warehouse and data mining", Communications of the ACM, Vol-36 (11), ISSN:0001-0782, pp-49-50 November (1996).

[104] Wani S P, Rupela O P and Lee, KK, "*Sustainable agriculture in the semi-arid tropics through biological nitrogen fixation in grain legumes*", Plant and Soil 174:129-49, (1995).

[105] William S. Cleveland, "*The Elements of Graphing Data*", revised, Hobart Press, (1994) Howard Wainer, Visual Revelations, Copernicus, (1997).

[106] Yadav, A.K., Raychaudhary, S. "Biofertlizers as Agro-Inputs-Past, Present and Future Perspectives", In A.K. Yadav, S. R. Chaudhary, and N. C. Talukdar, eds. Biotechnology in Sustainable and Organic Farming-Scope and Potential. New Delhi: Shree Publishers and Distributors. Chp. 1, p.1-17, (2004).

[107] Yu Zhao, Q. I. Guoqiang, "*Application of data warehouse in decision support system of rice cultivation management*",

Journal of Northeast Agricultural University, Vol. 37, No. 4, pp.557-562, (2006).

[108] Zhang X, Lesser V, Podorozhny R, "*Multi-Dimensional, MultiStep Negotiation for Task Allocation in a Cooperative System*", Autonomous Agents and Multi-Agent Sytems, 10, pp.5-40, (2005).

Web References

1. http://delnet.in
2. http://www.ifpri.org
3. http://www.twocrows.com
4. https://eands.dacnet.nic.in
5. https://krishi.icar.gov.in
6. https://ndl.iitkgp.ac.in
7. https://www.swayamprabha.gov.in
8. www.data-mine.com
9. www.kdnuggets.com
10. www.ncap.res.in
11. www.the-data-mine.com

■■■

Made in the USA
Monee, IL
04 November 2026

67381279R10134